Man

G000093890

Institutional Leases in the 21st Century

Chris Edwards and Paul Krendel

2007

EG Books

A division of Reed Business Information

Estates Gazette
1 Procter Street, London WC1V 6EU

ISBN 0 7282 0509 2
ISBN 978 0 7282 0509 3

Typeset in Palatino 10/12 by Amy Boyle, Rochester
Printed by Progress Press, Malta

Contents

Foreword

I have known Chris Edwards for many years, having worked with him in the very early stages of his conveyancing career. He has been an outstanding lawyer throughout his career and has always been at the forefront of understanding and implementing changes in property law as they emerged either through legislation or case law; and furthermore he has been involved throughout his career both from the landlord and the tenant perspective.

It is not surprising therefore that this piece of work which has been co-authored by Chris with Paul Krendel, whom he has known professionally for about 25 years, is outstanding and is essential reading for everyone from experienced conveyancers at one end to private individuals who may be contemplating entering the world of commercial leases for the first time. It provides at the same time a simple guide for those who want to have a basic understanding of the rudiments of landlord and tenant law but goes into sufficient detail for it to be a seminal piece of work for those who are deeply involved in the process.

The landscape for landlords and tenants is ever changing and the relationships between landlords and tenants have changed dramatically in recent years to the point where even the terminology has changed and we now refer to the less feudal owners and occupiers. It is hugely important therefore to have a publication such as this which provides an overview of the current situation as well as giving some background to the evolution of Landlord and Tenant law.

I am sure this will become essential reading for everyone involved in this complex area of legislation and case law and I congratulate Chris and Paul and their initiative.

Ian Coull
Chief Executive, Slough Estates

Dedication

For Michael Max and Silas Krendel

Evolution of the "Institutional" Lease

Institutions and property

Most of the cities in the UK have undergone significant redevelopment over the last 50 years. Take Croydon. Immediately after the Second World War it was a typical local centre with theatres, cinemas, pubs and a variety of local retailers occupying both department stores and small shops in its High Street. By the 1960s that was changing. A "Manhattan" skyline of office towers and (by 1970) the development of Whitgift School's playing fields into a large, uncovered shopping centre transformed the town. Why? There are, of course, many reasons — the willingness of the local authority to grant planning permission for such development, the location (a convenient staging post between green Surrey and the London conurbation) and excellent communications — but crucial to the process was the contract by which these buildings were let to their occupiers, the subject of this book: the institutional lease. The new buildings were sometimes owner/occupied (eg Taberner House, the new town hall). The majority, however, were developed and owned by investors — the "institutions" whose preferred form of lease dominated market thinking for the next 30 years and, in some ways, still does.

Interestingly, having transformed its traditional town centre in the 1960s, Croydon repeated the trick with the industrial land on the Purley Way in the 1980s and 1990s. This industry had itself mushroomed around Croydon Airport (which was London's airport until the Second World War and where Chamberlain waved his post-Munich piece of paper in one of the less accurate pieces of political soothsaying of the 20th century). The Purley Way was built as Croydon's bypass and the good road connections facilitated the growth of a number of industries; Mullard valves (later acquired by the Dutch electrical giant Philips), Trojan bubble cars, Sleepeezee beds, Payne's Poppets, to name but a few. With hindsight it is clear that valves and bubble cars were not industries that were likely to be in it for the long haul, even if they wished it otherwise. The early 1980s recession speeded the closure of marginally profitable manufacturing in the UK and land became available on the Purley Way. A giant Sainsbury's, IKEA (retaining the two chimneystacks of Croydon B power station on whose site it was erected), Currys, PC World, Toys'R'Us, B&Q, Mothercare, Comet, Habitat — all are to be found now along the alleged bypass which, at the weekends particularly, is now thronged with cars containing shoppers, and has had to be remodelled accordingly. By the mid 1990s Purley Way had become, as memorably described by a leading developer of retail warehousing at the time, "the Oxford Street of retail warehousing". And, as with that famous thoroughfare, the dominant landowners are the institutions who (or whose appetite for investment in commercial property) transformed the centre of Croydon 30 years ago.

Who are these institutions whose assessment of the terms of an occupational lease is crucial to the marketability and value of an investment? Some of them are large quoted property companies, such as Land Securities. Often, they are insurance companies and pension funds — investors with "real" liabilities where the obligation to pay must take into account inflation. These institutions may be contrasted with general insurers who may become liable to pay substantial amounts in the short term. They will invest in assets that can readily be converted into cash, for example equities and gilts, and leave part of their assets in cash for precisely that reason. Pension liabilities are closely related to inflation so the assets of choice for pension providers are index-linked gilts, equities and, of course, property. These are the assets which, historically, have maintained their real value over periods of high inflation. As we shall see, in the case of property investment, the inflation risk is borne by the tenant through the concept of rent review.

Property as an investment in comparison with bonds and equities

Equities

Historically, equities have out-performed other asset classes over time. They have a higher expected return than government bonds (reflecting their higher risk). They offer protection against inflation. Investment in overseas equities enables the investor to spread its risk. Emerging markets and unquoted equities offer high returns albeit for higher risks including illiquidity. The main component of their return is usually via capital gains rather than dividend income.

Bonds

These are interest-bearing certificates where the issuer undertakes to pay a stated rate of interest pending repayment of the capital sum originally invested at a future fixed date. They are issued by governments, local authorities and corporates. UK government bonds ("gilts") provide a risk-free income stream which is the benchmark against which other investments are judged. They are predominantly an income play, although there is a potential for capital gains (or losses) if they are sold before maturity

Property

In the UK, property as an investment usually means commercial property. The main sectors are retail, offices and industrials, although leisure and residential investments are increasingly available. The price of commercial property represents both the current rent and potential growth in rental income through periodic rent reviews, usually five-yearly. Those reviews should be capable of providing a hedge against inflation in the long term. The key attributes of property are:

- diversification benefits and a hedge against inflation
- stable income returns with cashflow advantages through rents being paid quarterly in advance
- expected returns competitive with equities and gilts (at the time of writing, property has been the best-performing asset class over the last one, three and five years)

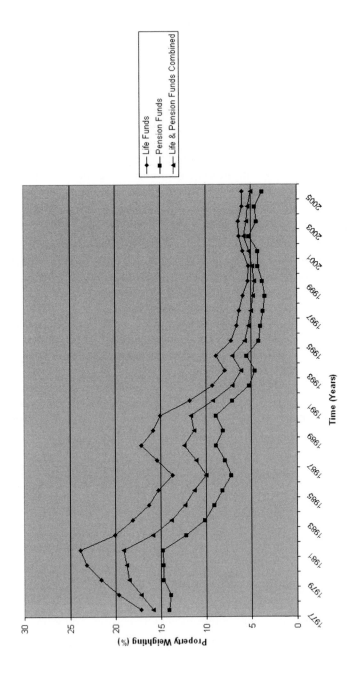

Figure 1.1 Property weighting over time

- relative illiquidity with high dealing and management costs
- poor information about transactions in the market
- downside risk limited by the institutional lease.

The income component of property's return is less risky than that of equities because a landlord's claim for rent ranks before that of a shareholder for company dividends. Companies may reduce or omit dividends but are obliged to pay rent as a prior charge. If a tenant goes into liquidation, a property may be relet.

Property returns indices generally show a low or negative correlation with those of equities or bonds. This offers a valuable diversifying asset for investors with "real" liabilities, yet, as we see from Fig 1.1, pension fund allocations to property have declined markedly over the last 25 years. A key factor is undoubtedly the perceived illiquidity of property as an investment. However, it is not unknown for large equity stakes in property companies to be more difficult to sell than large properties. Nevertheless the perception of equities as a more liquid investment than direct property remains. Yet property offers portfolio diversification with inflation-resistant qualities. This was traditionally achieved through the institutional lease which was as much a financial instrument as a contract for occupation. It offered the landlord a high-quality cashflow predicated on the following:

- full repairing and insuring liability placed on the tenant
- upward-only rent reviews offered a certain, inflation-hedged income
- priority over most other creditors for recovery of rent
- long lease terms (20 years+) offered longevity of income
- privity of contract — commonly known as "first tenant liability" — enabled the landlord to pursue former tenants where the current one defaulted.

Today, however, the picture is somewhat different.

A comparison of institutional lease provisions in 1983 and 2006

Table 1.1 sets out some of the main differences between an institutional lease granted in 1983 and one granted in 2006.

Table 1.1

Lease provision	1983	2006
Term	Usually 25 years. For emerging property types (eg pre-1985 retail warehousing) the term could be as long as 35 years	Unlikely (except for prime City offices or retail warehousing) to exceed 15 years, possibly with a tenant's break option at year 5 and/or 10
Tenant liability	The first tenant remains liable for the whole term by virtue of privity of contract. Subsequent assignees were usually required to give the landlord a direct covenant in the same terms, enabling the landlord to select from a range of defendants in default situations	Since 1 January 1996 the Landlord and Tenant (Covenants) Act 1995 saves the original tenant from continuing liability (usually) after its permitted assignee has itself made a permitted assignment of the lease. Direct covenants with the landlord are unenforceable
Rent review	Usually 5 yearly (shorter periods had been briefly fashionable in the high-inflation years of the late 1970s) to open market rent. The notional lease would be assumed to be on the same terms as the original. Disregards of rent, occupation and goodwill.	Still 5 yearly to open market rent. The lease may well make assumptions about the term length in the notional lease but the same disregards will usually apply Rent reviews still upward only with few exceptions
Break options	Virtually unheard of. Where they were included, their operation was sometimes conditional on the tenant having observed and performed all its obligations under the lease. This led to some high-stakes litigation in the slump of the early 1990s	Commonly found and usually unconditional. Valuers are coming to grips with their impact on capital values

Lease provision	1983	2006
Insurance	A well-advised tenant would insist on a full list of risks to be insured by the landlord at the tenant's cost. Isolated incidents of terrorism in mainland UK meant that this, too, was usually included in the list, often at the tenant's behest. This was a potential trap by 1995	A full list of insured risks was still a feature but provisions to apportion risk where an event could not be insured (terrorism after the Canary Wharf bombing, flooding in parts of Kent) were introduced — not always with a great deal of thought as to the nature and location of the premises or the reality of the landlord's and tenant's respective interests in the property
Other terms	All the other provisions of the lease (eg service charge) were designed with one aim in mind — to deliver a "clear" rent to the landlord by making the tenant pay for everything	While the 1983 philosophy is still popular with some landlords, the Government-sponsored Commercial Leases Code of Practice (originally introduced in 1995 and relaunched in 2000) sought to encourage landlords to consider more flexible lease negotiations than hitherto. The Code is reviewed in Chapter 10.

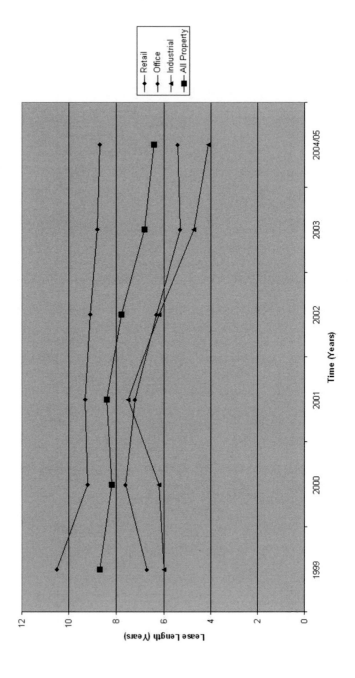

Figure 1.2 Correlation of lease length and property type

Term

In the 1980s it was understood that investment grade property would be let on 25 year terms. Indeed, some types of property where the fit-out costs were high (hotels, pubs) or where there were no ready comparables (retail warehousing which was in its infancy) were let for longer terms.

The IPD/BPF lease review is based on the evidence of over 83,000 tenancies. It shows a declining average lease term in all sectors during the 1990s (Fig 1.2). Will this trend continue? Certainly, external factors — the Land Registration Act 2002, Stamp Duty Land Tax and proposed changes to accounting standards, all considered in Chapter 9 — militate in favour of shorter terms.

Landlords maintain that tenants want longer leases to facilitate the writing off of their fit-out costs. It is instructive to consider the table, though. It seems to support the proposition that, usually, tenants will take shorter leases if they can. The trend started in the tenants' market of the early 1990s when it became possible to negotiate shorter leases. The lessons from overseas may also have been learned. Most tenants today even for the largest buildings wish to retain flexibility. Occupational requirements have reduced with technological advancement and tenants do not wish to commit themselves to buildings for 25 year periods where the property in question has become outdated and obsolete after 15 years.

One predicted downside of shorter lease terms is their effect on development. Long leases and the income they generate for developers and lenders have historically facilitated development in the UK. As we enter an era of ever shorter leases, how will development be financed?

Development finance

Development finance can take many forms and is beyond the scope of this book; suffice it to say that there are two kinds of development — prelet, where a tenant is signed up before the building is constructed, and speculative, where the building is built without a tenant being identified and committed. Prelets, unsurprisingly, are usually on "institutional" terms, either because the developer is itself an institution, or is being funded by one so must bow to that institution's requirements as to the form of lease. Speculative development in the late 1980s contributed significantly to the changes in lease terms in the

1990s. Sudden rises in interest rates, the vast overhang in available space and the subsequent insolvency of the developers who built it left financial institutions, especially the clearing banks, as major landowners and prospective landlords. In that situation, priority is given to cutting losses rather than creating institutional-type investments. The tenant's market which ensued saw the tenant gain ground in several areas (outlined in the table above) which, we suggest, will never be recovered by the landlords, institutional or otherwise. Today, speculative development is relatively unusual and although there is once more a surplus of space in some locations (eg City offices) it is clear that the banks will not end up as unwilling participants in the property market to anything like the extent of the early 1990s.

The financing of standing investments has always been a relatively straightforward matter for lenders where the property is let to a satisfactory covenant on institutional terms with at least ten years of the lease term unexpired. The market has responded with enthusiasm where such an investment is offered as security. However, the issues raised by vacant properties or properties let on shorter leases are more difficult for a prospective lender to address with the result that finance for such properties is more difficult to arrange.

The capital value of an investment property and its ability to service debt are both predicated on cashflow from the property and the likelihood of it being relet if the tenant departs. Lenders concern themselves with covenant status — simply, the perceived ability of the tenant to pay the rent — in a conventional loan. With shorter leases, the issues for the lender are the prospect of reletting, either to the current tenant or to another occupier, and the value of the property with vacant possession. These issues tend to increase the importance of the attributes of the property (location, specification) and reduce the emphasis on covenant status. The role of the valuer in helping the lender to form its view about the amount and terms of a loan in these circumstances is crucial.

Investment trends

Although pension funds and life companies are big investors in the institutional property market, they drive the equities market. As we saw in Fig 1.1, over the past 30 years they have moved away from direct property investment. There are a number of reasons for this. Pension fund management is usually outsourced. This leads to performance being driven by the fund manager's fear of losing his mandate. The pressure to be "doing something" militates in favour of equities and bonds as more trading puts a premium on marketability and liquidity. Key issues for the fund managers are information, lot size and the unavailability of international exposure. It is beyond the scope of this book to go into great detail on these issues. One solution which has commended itself to institutional investors over the last five or ten years has been the indirect investment offered by collective investment schemes, particularly limited partnerships. These are tax transparent and offer smaller funds the opportunity to invest with others in property types which would not be available to them as direct investments — for example, shopping centres. Unfortunately there is no evidence that the liquidity issue is satisfactorily addressed by these schemes. Many do not allow an exit during their life (commonly seven years) and attempts to create a secondary market in participations in collective investment schemes have met with only limited success. Finally, the recent introduction of Stamp Duty Land Tax and its adverse effect on trading in partnership shares has forced limited partnerships offshore.

The Pension Fund Partnership's 1998 Property Audit which examined investment by UK pension funds in property assets did not make particularly heartening reading for the property industry. The average investment in direct property was 4.4% of the fund although this disguised a wide range of weightings among the respondents. Twenty three percent of them had no direct property investments and of those over half had never invested in property and 95% had no plans to invest or reinvest in the sector.

Life companies have a more complex investment profile because they offer a wide range of products. They have both interest rate risk and inflation risk to deal with. This leads them to invest in assets which grow in relation to inflation. Short-term performance is not an issue for them. Their commitment is for the long term. To this extent, at least, property is a suitable investment for them.

Repairs and Service Charges

Before we embark on an examination of the importance of a tenant's repairing covenant in an institutional lease it is appropriate to consider a tenant's common law liability to repair premises let to it. The common law imposes only two obligations on a tenant:

- to use the premises in a "tenant-like" manner. This implies obligations not to damage the premises deliberately or negligently. A yearly tenant must also keep the premises wind- and water-tight and make "tenantable repairs". There is a dearth of authority as to the nature of these obligations
- not to commit waste. An act of waste is, broadly, one that changes the nature of the premises.

As far as the landlord is concerned there is very little liability. In one case a landlord was held liable to maintain the structure of a building to enable the tenant to comply with its obligations expressed in the lease to repair the interior but this was an exceptional decision on its facts. The general rule is that, unless the landlord covenants specifically, there will be no liability.

Statute has not intervened to assist the tenant of commercial premises as it has in some residential or mixed-use situations. The landlord and tenant can basically make their own bargain and, if the landlord is an institution having the whip hand in negotiations, you will not be surprised to find that the tenant takes on the repairing obligation. And, as we shall see, the common law position bears as much resemblance to the obligations usually imposed by an institutional lease as a crofter's cottage does to Canary Wharf.

The basic position

Letting agents speak of an "FRI" lease. FRI means the full repairing and insuring terms so beloved of institutions and, indeed, all landlords (despite the introduction of the Commercial Lease Code as to which see Chapter 10). The repairing component of the tenant's obligations is vitally important. To achieve a "clear" rent, the landlord must either be free of any liability to keep the building in repair or fully reimbursed for the repairs it carries out. The tenant is handed that responsibility. If the lease is of the whole building, this is relatively straightforward. In a multi-occupied building or, perhaps, on a development of units with a private road and landscaping the structure is more complicated. As it will be impractical for management and legal reasons to demise away the common parts, main structure, roof and foundations, the landlord will agree to undertake repairs to those items and recover the cost through a service charge. We look at arrangements for service charges in more detail below. The key is to ensure that someone is liable to repair all elements of the building or the development. It is in neither party's interest that an item of repair is unallocated because, as we have seen, the common law is inapt to supply deficiencies. And, of course, having allocated all the responsibilities, the landlord will wish to ensure that the tenant pays all the costs.

The philosophical question

Tenants are instinctively troubled by full repairing liability. Not only do they have to pay to use the building (rent) but they assume an obligation to keep it — at least — in repair and, in extreme cases, to return it to the landlord in the same or substantially the same condition as it was at the outset of the lease. In the case of institutional leases granted for upwards of 20 years this can be extremely burdensome (a review of the law on dilapidations is outside the scope of this book).

Where there is a service charge without a sinking fund the tenant may time its ownership badly and be liable to contribute in the year when the landlord decides to decorate the exterior of the office block of which the tenant holds one floor, or renew the lifts. Yet when there is a sinking fund, the tenant may contribute for years towards lift renewal, perhaps suffering elderly or unreliable lifts in the meantime, but have assigned its lease and left the building before the new lifts are installed. The potential injustice from the tenant's perspective is clear and there are no simple solutions. We look at service charges and sinking funds later in this chapter.

How is the repairing obligation defined?

There are two elements to this question. First, what do we mean by "repair"? Second, what is to be repaired?

"Repair"

The obligation to repair can only arise in the context of disrepair. If premises are not out of repair there can be no obligation to repair them. If premises are let that are not in repair at the time of the letting and the tenant covenants to "put and keep" or even simply to "keep" them in repair, the tenant has an immediate liability to repair what has been demised. However, this liability does not require improvements to premises that are in their original state. Thus in one case concerning a tenant's obligation, the tenant could not be required to repair a defect in the original construction of its premises that led to the basement of those premises flooding where the flooding did not cause damage to the premises. In another case, where the landlord sought to recover the cost of replacing an electrical air-conditioning system that worked sufficiently well to discharge the landlord's obligation to keep it "in good working order and repair", the landlord failed. The plant that the landlord had replaced would have been reasonably acceptable to a prospective tenant of the building of which it had formed part. In a third case, the landlord could not require the tenant to carry out works in anticipation of the eventual failure of plant even though the plant would have to be repaired or possibly replaced in time. At the point where the landlord sought to force the tenant's hand, the plant was not "out of repair". Finally, there is no obligation to do works that are desirable but not actually required to put premises into repair.

No commentary on repair is complete without quoting the Master of the Rolls' statement in the case of *Proudfoot* v *Hart* (1890) 25 QBD 42:

> the state of repair necessary ... in Grosvenor Square would be wholly different from the state of repair ... in Spitalfields.

The aspiring residents of Spitalfields would hesitate to come to the same conclusion in the early years of the 21st century! This leads to another point. The covenants in a lease speak from the date on which the lease was entered. Where an underlease is granted many years after the headlease out of which it was created, the repairing obligation in the underlease may have a different liability attached to it from the identically worded one in the headlease, depending on whether the location of the premises had improved or deteriorated since the headlease was granted.

What is to be repaired?

We have seen that there will be no repairing obligations in a lease of commercial property beyond those set out in the lease. It is therefore necessary to consider the definition of the premises to be demised to see the extent of the tenant's obligations. If the lease is a lease of part, the interrelation with other parts of the building both as to ownership and repairing liability must be carefully organised.

Leases of whole

Where the lease is of a whole building the issues are somewhat simplified. The landlord will still need to consider if and how the tenant is to be liable to repair the landlord's fixtures and fittings and to include them within the demise for this purpose. Consideration will also need to be given to the inclusion of other parts of the land owned by the landlord and falling within the curtilage of the building to be occupied.

Leases of part

Here the situation is potentially more difficult. A well-drawn definition will include all internal surfaces and one half of all non-structural boundaries (walls and floors) separating the demised premises from other parts of the building. The landlord will retain the structure, roof, foundations and common parts and services (eg lifts) and recover the cost of repair of those elements through a service charge (see below).

Structural repairs

In a lease of part, the landlord will often covenant to maintain the "structure" or "main structure" of the building. The expression has been popular for many years but avoided legal or judicial definition until 1958. The distinction was drawn between structural and decorative repairs. This led to some odd outcomes and, in one case, works involving replacement of about 3% of some roof tiles were held to be structural repairs on that basis. The expression "main structure" may be more limited and has been held not to apply to horizontal divisions (in one case, the floor separating the ground floor of a building from the basement) or the decorative finish of a roof terrace.

Fixtures and fittings

While a discussion of the law relating to fixtures and fittings is beyond the scope of this book, institutional landlords are astute to impose repairing obligations on tenants in relation to fixtures and fittings within the demise. It is clear that anything affixed to the premises during the term that becomes a landlord's fixture will be within the covenant. It is also likely that tenant's fixtures attached to the premises would fall within this category. Well-advised tenants would usually exclude tenant's fixtures and fittings from the repairing obligation. A similar analysis applies to improvements. Improvements to and forming part of the demised premises are usually included because they become part of the land. If a tenant makes improvements that it does not wish to become obliged to keep in repair, the obligation (or lack of it) should be defined in any necessary licence for alterations. There could be some overlap here with the assumptions and disregards on rent review.

Repair versus renewal

We have already seen that the repairing obligation speaks from the date of the lease. An institutional lease of newly-constructed premises is unlikely to confine the tenant's obligation just to "repair". The obligation will commonly include requirements to "replace, rebuild, renew". Indeed, for a while it was fashionable to include a specific obligation to put right latent or inherent defects in the premises but doing so ran the risk of the rent being depressed on review to reflect the onerous nature of the obligation (see Chapter 5). Even so, a tenant who has escaped with an obligation confined to "repair" may not be entirely absolved from liability when things go seriously wrong with its premises. Much judicial wit has been expended in distinguishing works of repair from other kinds of works, particularly improvement and renewal. As there are as many possible decisions as there are factual situations leading to a dispute, it is impossible to offer a certain outcome in any case. Some useful principles can nevertheless be distilled from judicial pronouncements over the years:

- repair is restoration by renewal or replacement of subsidiary parts of a whole (1911)
- if the work which was done is the provision of something new for the benefit of the occupier, that ... is an improvement; but if it is only the replacement of something already there, which has become dilapidated or worn out, then ... it comes within the category of repairs (1956)
- [the correct approach is] to look at the particular building, to look at the state which it is in at the date of the lease, to look at the precise terms of the lease, and then to come to a conclusion as to whether, on a fair interpretation of those terms in relation to that state, the requisite works can fairly be termed repair. However large the covenant it must not be looked at *in vacuo* (1970).

"Large" covenants are a hallmark of the institutional lease and the inclusion of obligations to renew, rebuild, etc. largely stymied attempts by tenants to avoid repairing liabilities, at least at their high watermark — probably in the early to mid-1980s.

Inherent defects

It was around this time that a liability to repair inherent defects became a feature of institutional leases. An inherent defect may be defined as one that manifests itself eventually as a result of substandard design or construction at the outset. Judicial moves to place inherent defects outside the repairing obligation, as analogous to "replacement" or "improvements" were dealt a body blow by the decision in *Ravenseft* v *Davstone* in 1980. In this case the cladding on a block of flats began to distort because the initial construction had omitted expansion joints. The only (and very expensive) remedy was to remove the cladding and insert expansion joints before attaching the cladding to the building once more. This was held to be a "repair". The courts have enunciated the following principles in the development of case law about inherent defects:

- in 1970 it was held that a tenant cannot be required to do repairs that are so extensive that they result in the landlord getting back a different property from the one it originally let
- in 1980 *Ravenseft* obliged the tenant to repair defects causing damage so that the damage itself could be remedied "for once and for all"
- this message was reinforced in 1986 by a decision that some improvement may be required in the course of repairing damage so that the damage would not recur
- but in 1987 tenants received some small relief from the decision that an inherent defect that did not cause physical damage did not amount to disrepair and so could not fall within the tenant's repairing liability.

Naturally, institutional landlords would not leave such a possibility to the chance that a more sympathetic judge or a slightly different set of facts would absolve the tenant from liability. The tenant would have to find another solution which it did through the development of collateral warranties. These are contractual obligations from the contractor and professional team involved in the design and construction of a new building. Those obligations were, of course, also owed to the institutional landlord (or perhaps the developer from whom the institution had itself contracted to buy the building on completion and letting in which case the institutional landlord would itself have a right of action usually arising under similar warranties). There are a couple of points to note here. First — unsurprisingly — this arrangement gives the institutional landlord two bites of the cherry. It can pursue its tenant or it can pursue the contractor/professional team. If it pursues the tenant the tenant can in turn pursue the contractor/professional team to the extent of the warranties. Second, the writers were formerly unaware of any reported cases of tenants enforcing or seeking to enforce their rights under contractual warranties. However, in a case reported in early 2006, the court found for technical reasons that the liability of the contractor under the warranty could not exceed its liability under the main contract, so the tenant's claim was defeated. While the warranty contained a relatively unusual limitation of the contractor's liability, it may not be assumed that the existence of a collateral warranty will necessarily provide the tenant with the comfort it needs in the face of a claim by its landlord.

As noted elsewhere, a repairing obligation that renders a tenant liable to repair inherent defects can have a depressive effect on the rent achievable at review.

Ways in which a tenant might limit its obligation

We will see in Chapter 10 how tenants have improved their position in negotiations since the publication of the Commercial Lease Code. Repairs have always been a key concern of tenants under institutional leases but help has been forthcoming from or negotiated in the following areas.

Statutory protection

The Leasehold Property (Repairs) Act 1938 applies to a covenant to repair commercial (and, indeed, any non-agricultural) premises contained in a lease originally granted for more than seven years of which at least three are still to run. Where this Act applies, a landlord seeking to forfeit or (more likely) recover damages for breach of a repairing covenant must first serve on the tenant a schedule of dilapidations under section 146 of the Law of Property Act 1925 accompanied by a statement of the tenant's rights under the 1938 Act. These enable the tenant to serve notice on the landlord within 28 days requiring the landlord to obtain the leave of the court before seeking to enforce its remedy.

Schedules of condition

We saw that in lettings of new buildings a tenant who might otherwise be obliged to maintain a new building to a high standard can lay off some of its risk to the contractor and professional team responsible for the initial construction of the premises. Where the premises are second hand it is not always possible for the landlord to put the same burden on its tenant. In these cases, the tenant can negotiate a limitation of its liability to keeping the premises in no worse state or condition than they were at the date of the lease. That state or condition is evidenced by a written schedule of condition, more often than not since the advent of digital photography with an extensive gallery of pictures evidencing the actual repair of the premises.

"Damage by insured risks excepted"

Where the landlord insures — which is usual in the case of institutional leases — the tenant's repairing obligation should be qualified with these or equivalent words to exclude the possibility of the insurer seeking to enforce the repairing obligation against the tenant after paying out the landlord. The qualification assumes that the insured risks will have been defined, which they will have been in a well-drafted lease.

Damage by uninsured risks

In recent years the issue of damage by an uninsured risk has come to the fore because of terrorist attacks. The problem then is how to apportion liability between the landlord and the tenant in a fair way. This question attracted a lot of interest in the mid-1990s but seems to have moved out of the spotlight recently, perhaps because insurers are once more offering cover for this risk. A solution that does not oblige the tenant to repair but gives the landlord the choice of doing so on notice or determining the lease has found most favour. Most institutional leases impose an obligation on the landlord to insure against a list of defined risks without exceptions but with all the costs being passed on to the tenant. This is potentially very expensive. Assignees of institutional leases granted before 1992 should always check the insurance provisions carefully, not only to establish what risks are insured against but also to see whether the landlord is obliged to insure them regardless of cost — a cost that will inevitably be passed on to the tenant.

New buildings

We saw above that a tenant can sometimes obtain contractual warranties from the contractor/professional team where it takes a lease of newly-constructed premises. Other solutions for the tenant might be:

- the exclusion of inherent defects from the repairing obligation. This will be difficult to negotiate where the landlord is offering contractual warranties
- decennial insurance covering defects arising through failings in the design or construction that manifest themselves in the first ten

years of the term. It is a fact that most defects in new buildings appear in the first three to five years after completion. This insurance is usually put in place by the landlord as the insurer will wish to inspect the building during the construction process to ensure that there are no obvious problems with the building process.

The principles of these exclusions apply equally to situations where the tenant assumes a full repairing liability or the cost of repair is recovered through a service charge levied by the landlord to cover its expenditure on the maintenance of the building.

Service charges

Leases of multi-let commercial buildings (usually office blocks) or developments (shopping centres, business parks, industrial estates) invariably contain service charge provisions. The landlord recovers the cost of maintaining unlet structures, common parts and the provision of other services through this mechanism. Thirty years ago the charge might have been a fixed annual sum or subject to a primitive form of indexation. The service charge for an office building in Birmingham was famously linked to the price of oil in the late 1960s. By the mid-1970s the rise in the price of oil had outstripped almost any other measure of inflation that might have been chosen. The tenants still had to pay. This episode could in theory be repeated because statute has never intervened in the service charge arrangements between landlord and tenant in a lease of commercial premises. The position is different in the residential sector. The application of common law principles to service charge disputes can ameliorate the tenant's position, as we shall see.

What may be recovered?
Scope of services

We saw earlier in this chapter that the landlord will usually retain responsibility for the structure and exterior of its building or development. The actual scope will be defined in the lease. As a matter of principle the landlord will want to be able to recover all its expenditure from the tenant. The tenant will try to limit the recoverable costs to items that are necessary for or appropriate to the development and the tenant's interest in it, whilst precluding the landlord from recovering costs that are properly for the landlord's account. The standard of provision of services is not subject to a "lowest common denominator" test — the landlord can choose a standard that is reasonable for the service that is being provided in the context of the wording of the lease and what someone who owned the building and would have to bear the cost itself might consider reasonable. A recent case opened the door slightly to a tenant's successful objection nevertheless. It was held that the landlord must take into account its tenant's limited interest in the premises, in effect allowing the term of the tenant's lease to be the determining factor in allocating cost.

Capital expenditure

The logical outcome of a "clear" lease is that the tenant either keeps the building as good as new itself or pays for the landlord to do so. We consider sinking funds below, but the issue here is whether the tenant should agree to pay for capital items as a matter of principle. The courts have intervened to strike a balance between the landlord's and the tenant's interests in the property (see previous paragraph) and the tenant should be alert to ensure that it does not find itself with a heavy service charge bill in one year of what might be a relatively short commitment as occupier. The obligation to reimburse capital expenditure may be hidden in the wording of the service. In one case, the "total cost" of providing lifts and hot water was held to include capital costs of providing those services. The tenant had to pay.

Interest

An institutional lease will commonly seek to entitle the landlord to recover the cost of interest charged to it on money borrowed to pay the cost of services. This may be reasonable where such borrowing actually occurs and is certified as part of the annual charge. More sinister for the tenant, but also seen, is a provision entitling the landlord to make a charge for notional interest foregone by the landlord where it has paid for the services without borrowing. The applicable period during which this "interest" runs will usually be from the date of expenditure to the date of demand. A properly advised tenant will argue in negotiations that owning property involves some costs which cannot reasonably be passed on. It may be thought that the notional interest provision is one of the high watermarks in the quest for a "clear" lease. Interest will also be incurred by the tenant in the usual way where it is late in paying the service charge. The point to note here is that such interest will be for the landlord and need not be credited to the service charge account.

Professional fees

Very clear words indeed will be required to enable the landlord to recover fees beyond basic management. Attempts by the landlord to recover solicitors' fees for proceedings relating to unpaid rent or service charges (either debt recovery or forfeiture) have met with short shrift in the courts. Well-drawn leases will require the defaulting tenant to pay the landlord's costs incurred in such an action and it may be a reluctance to give the landlord a second bite of the cherry that informs the courts' reaction to attempts to recover through the service charge. Indeed, the courts will imply a duty on the landlord to recover those costs from the tenant in breach before attempting to recover through the service charge where the service charge would otherwise have permitted recovery. Professional fees in relation to activities that are not themselves in the definition of services will not be recoverable either.

"Sweeping up"

An institutional lease will usually contain a clause in the service charge provisions purporting to entitle the landlord to recover any expenses not otherwise recoverable under any other head of charge. The courts interpret "sweeping up" clauses strictly and the law reports are littered with cases where the landlord has failed to recover the cost of what were, on an objective analysis, improvements that went beyond the concept of "services". The landlord will attempt to secure his position by including a declaration that the parties agree that the landlord is to be able to recover all costs and expenses relating to the building. The tenant has a different agenda and will wish to confine any non-specific recovery to services:

- reasonably calculated to benefit tenants "generally"
- reasonably and properly provided
- reasonably expected to be for the benefit of the tenants in the building.

The power of the landlord in, for example, a popular shopping centre is usually manifested by reacting to a tenant's (apparently) reasonable proposals along these lines by saying that leases have already been granted containing certain service charge provisions and that the landlord is not in the business of customising the service charge provisions for each tenant.

Calculation and payment
Methods of calculation

The lease will specify the tenant's obligation to pay its share of the service charge. But how is that share defined? There are various possibilities, some of which might favour the tenant and some the landlord. We may reasonably assume that the tenant, having sought to limit the number and type of services to which it is liable to contribute, will also wish to limit its share of the cost of those services. For the institutional landlord, the name of the game is full recovery. The possibilities are considered in that light.

- Fixed sum (eg £1,000 pa). This has the advantage of alleviating the landlord's management expense (no calculation is necessary). It is certain, but almost certainly wrong! This will be a problem for the tenant if it is too high and for the landlord if it is too low. Despite that, a fixed service charge or even a service charge-inclusive rent is coming to the fore as lease lengths fall, especially in developments where there are short leases and a turnover of tenants such as serviced offices or short-term industrial premises (see Chapter 11).
- Fixed proportion (eg 12.5%). Some calculation is required, but the outcome is certain while not necessarily fair between tenants, particularly where circumstances change such as the landlord acquiring more premises to add to the original. A proportion that can be varied in the light of future changes can work fairly for both parties.
- Fair proportion. Must this be "fair" (or "reasonable" or even "proper")? The landlord may not think so if litigation ensues. In one case the court held that a tenant need only contribute a small amount to the cost of a roof repair because the new roof would last about ten times as long as the unexpired residue of the tenant's lease at the date of demand of the contribution.

- Proportion to be found by calculation. The method of calculation can be crucial here. There are various possibilities. For example, the proportion that the floor area of the premises bears to the floor area of the whole; in this case the landlord will have to define the "whole" carefully as including unlettable parts (eg corridors) in the total will inevitably lead to a shortfall. Sophisticated methods of weighting floor areas exist to even out the use of services in complex multi-let situations such as shopping centres, usually at the behest of the anchor or larger tenants because their presence is of benefit to the occupiers of the smaller units. A simpler method is to allocate expenditure by reference to the number of units in the development. Alternatively, the calculation could be made on the rateable value of the premises as a proportion of the rateable values of all the lettable premises in the development. This can be complicated by rate revaluations — in one case it was held that the calculation must be based on the rateable values at the date the landlord incurs the service charge expenditure, not the date when it calculates the bill. None of these methods really addresses the tenant's concern that it may find itself subsidising another tenant whose use of the services or some of them is greater than the average.

The tenant's obligation

Institutional landlords like the tenant to covenant to pay the service charge as rent so that the landlord will have the greatest range of remedies if the tenant does not pay. These would include distress. A well-advised tenant will run the argument that, as service charge arrangements are potential sources of dispute, the landlord should not have draconian remedies available to it where the tenant wished to argue. Unless the service charge is expressly reserved as rent it will not be rent so that, for example, the landlord would have to serve a notice under section 146 of the Law of Property Act 1925 as a prelude to forfeiting the lease if it were not paid. All institutional leases would reserve service charge as rent. Those that do not were probably negotiated during one of the periods in the economic cycle where the tenant had the upper hand.

Timing of payments

The usual arrangements are set out below.

- If the services are simple and irregular, the lease may oblige the tenant to pay its share to the landlord on an "as and when" basis.
- If the services are more substantial and/or continuous (eg lift maintenance in an office block or cleaning the common parts of a shopping centre) the landlord will want the tenant to make on-account payments during the service charge year with a balancing payment or credit against the following year's expenditure at the end of that year. Otherwise the landlord will find itself funding all expenditure until after the end of the service charge year which, understandably, it will not wish to do. Two cases in the late 1970s suggested that a "reasonableness" test might be implied into the calculation of the on-account payment. The Court of Appeal was unwilling to infer from a lease that the landlord could demand payments anticipating the highest possible standards of repair.
- If the arrangements are such that periodic renewal of capital items by the landlord is funded by the tenants, as well as on-account payments the tenant will be obliged to contribute to a sinking fund designed to anticipate large capital expenditure by building up a reserve against it. We look at sinking funds in more detail below.

Landlord's obligations

Institutional leases impose many obligations on the tenant but, at least at first draft stage, are often light on landlord's obligations. There is no implied correlation between the long list of services (including a "sweeping up" clause) of which the landlord can recover the cost and the landlord's obligation to provide those services. The tenant should therefore seek covenants from the landlord:

- to provide services (the landlord will often compromise with an agreement to use all reasonable endeavours to do so)
- to fund the service charge for vacant units within the development
- to hold advance payments and sinking funds on trust for the tenants
- to provide services "in accordance with principles of good estate management". This wording is often seen but has yet to be

judicially defined. Landlords may resist it for that reason, or take a chance on the vagueness of the phrase making it difficult for an accusation of non-compliance — short of obvious failure — to "stick". In particular, it will not oblige the landlord to adopt the cheapest service or method of repair.

- to enforce rights it may have against contractors or professionals to require them to repair the main structure and the common parts etc before having recourse to the service charge payers.

Even if the landlord agrees to some or all of these obligations, it will usually try to limit its exposure by excluding

- liability for failure to provide services for reasons beyond its control
- liability for loss or damage caused by the failure of any plant or machinery or the interruption of any services
- obligations to extend the scope of the services (although landlords usually like to be able to do so at their discretion).

Sinking funds

The modern trend towards shorter leases may not be the final nail in the coffin of the sinking fund, but it is certainly a contributory factor to their rarity. The theory is fine — the tenant makes an annual contribution towards major capital expenditure falling within the service charge (eg lift replacement) and which, if charged on an "as and when" basis would make for a very large bill one year. But the road to hell is paved with good intentions. Some sinking funds still exist, of course, but they can spell trouble for the landlord and injustice for the tenant for the following reasons:

- unless the fund is carefully constituted, it may be vulnerable to a claim by the landlord's creditors if the landlord becomes insolvent
- the long-term nature of the fund is such that the tenant may never see the benefit of the works towards the cost of which it has been required to contribute (and, if it does not, it has been held that the tenant has no right to require repayment even at the end of its lease)
- the tax position of the fund may be disadvantageous and complex. These issues are beyond the scope of this book; suffice it

to say that income and corporation taxes, capital allowances and (if the fund is held on trust) inheritance tax may all have to be taken into account in setting up and managing the fund.

"Service charges in commercial property — a guide to good practice"

Unlike residential property, where there is protection for the leaseholder enshrined in statute, there is no such regime for commercial premises. As we see elsewhere in this book, pressure from tenants and a desire to be seen to be reasonable from landlords (thereby avoiding the risk of legislation) has led to two publications of interest in this area.

- In 1986 a joint sub-committee of the Law Society and the RICS prepared a paper on model lease clauses.
- In 1994 a working party of representatives from a number of industry bodies (the BCO, the BCSC, the BPF, the BRC, the PMA, the RICS and the Shopping Centre Management Group) produced *Service Charges in Commercial Property — a guide to good practice* that can be read via *www.servicechargeguide.co.uk* . The guide is now in its second edition (August 2000). Apart from some interesting and realistic comments about the theory of service charges, the guide contains proforma reports and accounts the adoption of which would, it is thought, go some considerable way to allaying tenants' instinctive fears that they are the victims of opaque accounting techniques and decision making for which they, ultimately, must pay.

Valuation implications

Clear leases

If a commercial building is let to one tenant on a full repairing and insuring lease, this will have a neutral effect on the rental valuation at rent review, as the vast majority of buildings are now let on full repairing and insuring terms.

There may however be an adverse impact if for example the lease places an obligation on the tenant to re-build and the property is already an old building, as this might be perceived on rent review to be an onerous obligation. The prudent landlord will consider very carefully, before inserting a tenant's obligation to re-build in a lease of such a building. Words such as "renew" or "replace" are less onerous and may only relate to a component part of the building such as a passenger lift or an air conditioning installation.

Internal repairing leases

In a situation where a tenant takes a lease with only internal repairing and insuring obligations, excluding its liability to carry out external repairs or redecorations to the roof or structure of the property, this will undoubtedly have an impact on the rent to be assessed on a rent review. In a situation such as this, the tenant will have a favourable lease when compared to the majority of leases granted in the open market, and the landlord will expect to receive an additional rent to reflect the obligations which it has retained in dealing with the exterior repairs and decorations to the property, the roof and the structure. The amount of the addition in the rent will depend on the exact circumstances, but it is not unusual to find that an addition of between 2.5 and 10% will be made to the rent, in order to compensate the landlord for the less favourable lease terms. However, in order accurately to assess the quantum of additional rent, regard should be had to the actual costs and the likely repair works which are going to be needed to the exterior, roof or structure over the next five to 10 years.

Service charges

When dealing with a rent review of a multi-let building, where the tenants pay a service charge to the landlord, the payable amount per sq ft needs to be considered very carefully when undertaking a rent review of the accommodation, whether acting for landlord or tenant. If the service charge is either lower or higher than the normal payable amount for a building in a particular location, considering the standard of building and amenities that are offered, again there will need to be an adjustment made to the rental value; upwards, if for any reason the service charge is considered low, or downwards, if the service charge is on the high side effectively meaning that the tenant is having to pay a higher than normal amount for the service charge, which may have an impact on the rent which could be achieved if letting the accommodation on the open market.

Sinking funds

From a valuation perspective, this is something of a "double edged sword" because a prudent landlord, may consider that it is extremely wise to set up a sinking fund for the eventual replacement of items of capital expenditure, such as the roof, installation of a new passenger lift, or replacement of the air conditioning installation. This will, however, have the effect of increasing the level of service charge paid by the tenant, which may have a detrimental effect to the landlord on the rent which can be achieved for the property, either on first letting or upon rent review. The advantage of a sinking fund is that a large capital sum will have accrued over a period of years, so that the amount in question does not have to be suddenly found, when a major item of capital expenditure is required, but this can often have a negative effect on a tenant after contributing for many years to a sinking fund and then wishing to assign its lease, perhaps one or two years before the sinking fund monies are to be used for the replacement of the item in question. Sinking funds are a common item found within a service charge for a large multi-let building, but they are by no means universally used and as previously explained, there are advantages and disadvantages.

User and Alteration Clauses

In a well drafted institutional lease, there are two very important clauses which both landlord and tenant will wish to ensure are carefully drafted in order for the provisions to reflect the wishes of the parties. A balance needs to be found between the clauses having the proper degree of control for the landlord, against the possibility of the clauses being overly restrictive, which might have an impact on rent review or upon renewal of the business lease, where essentially the tenant will be entitled to a new lease upon similar terms and conditions to the old lease.

User provisions

Basic principles

If there are no restrictions on use in the lease or tenancy agreement then, as between the landlord and the tenant, there is no restriction on the use to which the tenant may put the demised premises. The tenant must, of course, observe any planning constraints, the common law rules against nuisance and similar unneighbourly behaviour and any other relevant statutory restrictions. However, most leases or tenancy agreements do contain restrictions on how the demised premises can be used, and the control is generally achieved by a clause containing a main covenant governing the permitted use and other clauses dealing with town planning, nuisance and additional or ancillary restrictions.

The form of the user covenant in a lease is a matter to be negotiated between the landlord and the tenant at the time the lease is being taken.

From the tenant's point of view, if an existing lease is being acquired, the user covenant should be carefully checked to see whether the proposed use is permitted and the extent to which the landlord's consent to a change in the use will be needed and, if needed, will be available.

From the landlord's point of view, the user covenant should be consistent with any restrictions on the landlord itself and should reflect the degree of control the landlord wishes to exercise over the tenant.

Positive covenants

Often words such as "to use" and "to keep ... open" are found within the user provisions and are positive obligations. They impose a positive duty, in the first case, to use the premises for a particular purpose or to which the use is restricted and, in the second instance, to keep the premises open for trade. The tenant will be in breach if he does not use the demised premises for the purpose specified or. as the case may be, keep them open.

Negative covenants

The words "not to use" impose a restriction on use and not a duty to use. They constitute a negative covenant. There are various examples of negative covenants, which fall outside the purpose of this book, and sometimes reference in the lease is made to town planning legislation, most typically the Town and Country Planning (Use Classes) Order 1987 (as amended).

A major change in planning legislation, might make a significant change to the range of uses to which a property can be put. One of the most telling examples of this was in the 1987 Use Classes Order, when the B1 business class was introduced, which at a stroke converted previous Class 3 light industrial uses under the Use Classes Order 1972 into business use classes. This largely caused a very substantial increase in rental value for this type of use, which overnight became classified in the same use class as office accommodation. We recall some commercial property space in Soho, which was substantially refurbished and reconstructed in the early 1980s for light industrial use and then by 1987 with the new Use Classes Order, suddenly became a much more attractive proposition, falling within the same use class as office accommodation within Soho, which attracted a much higher level of rent than light industrial accommodation had previously. Tenants were faced at the next rent reviews with a much more substantial increase in the rent than otherwise would have been the case.

Keep open provisions

Often, most frequently with shops, there is a "keep open" provision within the user clauses in the lease and typically this will be related to the usual business hours in the locality or perhaps within a covered shopping centre. The tenant should try to ensure that this obligation is qualified so as to allow closure for eg refurbishment, repair, annual holidays and in anticipation of a permitted assignment or underletting. Such covenants are enforceable but the courts have been reluctant to order specific performance to make tenants comply. In the right circumstances, the court will award damages against a tenant who has breached a positive covenant to trade and, only in exceptional circumstances, may be prepared to grant a mandatory injunction. Whether a use ancillary to the permitted use, amounts to a breach of covenant, is a question of degree. If an activity is generally ancillary, a breach is unlikely to have been committed. For example a covenant "not to use premises other than as a shop" will not be breached if part of the premises are used for storing items to be sold in the shop or for staff facilities.

Competition

Sometimes a landlord will covenant with a tenant not to let adjoining or nearby premises for a particular use in order to preclude competition with the tenant's business. Landlords are generally very unwilling to give tenants covenants against competition as such covenants may put the landlord in an exposed or dangerous situation. More common are clauses which restrict the tenant sometimes found in the alienation provisions, where the landlord may reserve the right to withhold consent to assignment on the grounds of quality, user mix, or particular trades and businesses. This may be to prevent the situation in a parade, where the tenant wishes to assign its lease to a user which is already found within the parade and this type of situation might create business difficulties for the original tenant using the shop for a particular purpose. This could also have an adverse effect on the landlord in the tenant's ability to pay the rent.

Changes of use

Absolute restriction

An example of an absolute restriction in a user clause is "not to use the demised premises other than as a shop for the sale of shoes and leather goods and for no other purpose". In such a situation, the tenant must try to negotiate a variation of the lease with the landlord which permits the change of use immediately required by the tenant and, ideally, admits the possibility of future changes of use (subject to licence). The less satisfactory result is that the landlord grants a licence for the change of use needed to meet the immediate need. The landlord may demand a premium or some other financial contribution as the price of his agreement to the change of use and modification of the terms of the absolute covenant.

Qualified restriction

The most usual example of this is "not without the landlord's prior written consent" to use the demised premises other than as "a shop for the sale of shoes and leather goods". There is no statutory implication that consent is not to be unreasonably withheld. In this type of qualified restriction, the landlord may not demand a premium for giving its consent, as opposed to the situation with an absolute restriction on use, where the landlord may demand a premium or some other financial contribution, as the price for obtaining his agreement to a change of use. Careful consideration of the provisions of section 19(3) of the Landlord and Tenant Act 1927 must be given in this type of situation.

"Consent not to be unreasonably withheld"

If these words appear after the mention of change of use not being permitted without the landlord's prior written consent, this has the effect of entitling the tenant to obtain the landlord's consent, where it is reasonable to do so. This can be a contentious area and each case will have to be decided on its own particular merits. Unlike applications for assignment or underletting (as to which see Chapter 4), the Landlord and Tenant Act 1988 does not apply to applications for licences for a change of use and there is no statutory duty on the landlord to give a consent to a change of use within a reasonable time or to show that any condition or refusal of consent was reasonable. The court will decide on the particular facts of the case, whether the landlord was indeed being unreasonable and this may extend to a particular long decision making process in dealing with the tenant's application for the change of use.

Estate management considerations

These may arise in a number of different ways for example:

(a) in a shopping parade, the landlord may want to prevent competition with its own trading activities or between its tenants
(b) in a retail park or speciality shopping centre, the landlord may want to ensure a satisfactory tenant mix for the success of the centre or to prevent or limit competition between tenants
(c) in a high-class office development, the landlord will want to attract tenants who will suit the image of the development.

Restrictions on flexibility and maximisation of rental income

There is a dichotomy here in that if the user provisions are too restrictive, this may well have an adverse effect on the landlord's maximisation of rental income. It is possible for the landlord to "have its cake and eat it" by having a restrictive user clause, but then in the rent review provisions, effectively ignore the actual user restrictions in the hypothetical lease where it is assumed the tenant has an unrestrictive user clause. A properly advised tenant will not accept this type of drafting at the beginning of the lease and therefore as tenants have become better advised and more sophisticated, it is unusual to find this type of situation in a commercial lease. An example of what can happen if a restrictive user clause is imported into the rent review provisions was the case of *Plinth Property Investments Ltd v Mott, Hay & Anderson* (1978) where the user covenant confined the use to that of a civil engineering business. If the lease had not contained this specific restriction, the market rent on review would have been £130,455 pa, but due to the restriction, the rent was fixed at £89,200 pa, being a reduction of 31.6%, to reflect the very limited market where effectively the only potential tenants for the office accommodation were civil engineers within an office building located in Croydon, South London.

Generally speaking, the more restrictive the user clause, the lower the rent achievable upon review is likely to be (and indeed upon letting). While a narrow user clause may save the tenant rent and reduce the return to the landlord it may, later in the term, have a "knock on" effect on the tenant's ability to deal with its interest either by way assignment or subletting. A tenant must give careful consideration not only to its present requirements, but also to its likely future requirements over the term of the lease.

Lease renewal

If the lease does not have the protection of the Landlord and Tenant Act 1954, the parties are able to agree whatever they wish upon renewal and even if there was a restrictive user clause in the previous lease, by agreement the parties upon a renewal may decide to make the user provisions unrestricted in the new term. If the lease is within the Landlord and Tenant Act 1954, the landlord and the tenant can agree the form of the user covenant in the new lease or, in default of agreement, the court will decide. The court will have regard to the current tenancy and all the relevant circumstances. Following *O'May* v *City of London Real Property Co Ltd* [1982] 1 EGLR 76, it has become an established principle that the parties will generally be entitled to a new lease upon similar terms and conditions to the old lease, meaning that if there is a restriction on use in the old lease, the court in the absence of agreement, unless there are important reasons to change the user clause, will impose a new lease with similar restrictions. A tenant with a restrictive user clause in an old lease, will have to consider very carefully, whether it is appropriate to have a new lease with similar restrictions, or accept the possible opening up of the user clause in the new lease, which although leading probably to a higher rent, will mean that it may be easier in the future to find an assignee or underlessee for the accommodation, should a disposal become relevant during the new lease.

Alterations and improvements
Control by landlords

The properly drafted institutional lease will provide for the landlord's strict control over the tenant's alterations, but again the well advised landlord will be careful to ensure that the controls are not overly restrictive, as this will have an impact on rent review.

Apart from obtaining the landlord's consent, some or all of the following permissions may also have to be obtained by the tenant:

- planning permission
- listed building consent
- building regulation approval
- obtaining or revision of fire certificate
- Disability Discrimination Act compliance.

It may also be necessary for notification of alteration works to be provided by the tenant to the landlord's insurers, possibly resulting in an additional premium. In obtaining the landlord's consent, there will normally be a landlord's licence prepared by the landlord's solicitors, and the costs of obtaining the landlord's approval and if necessary instructing solicitors to produce the licence, will be paid for by the tenant.

There are different types of alteration covenant, ranging from absolute prohibitions, to qualified covenants generally to the effect that the tenant shall make no alterations without the prior written consent of the landlord, to fully qualified covenants, stating that the landlord's consent shall not be unreasonably withheld. This is the most usual type of covenant and, unlike user clauses, if the covenant is qualified, there is an assumption under the Landlord and Tenant Act 1927 section 19(2), that the landlord's consent will not be unreasonably withheld unless the alterations are not improvements. Very often, in the case of office lettings, there will be a provision that the tenant may erect or dismantle non-structural internal partitions without the landlord's consent. The licence for alterations will normally contain a standard provision that the landlord is entitled to insist that the tenant reinstates the premises to their original condition at the end or sooner determination of the lease. Occasionally, this will not be enforced by the landlord where for example the alteration has created an improvement to the specification of the premises eg the installation of an air conditioning system to a building that was previously only centrally heated. It is unlikely that the landlord would insist that the tenant removed the air conditioning system at the end of the lease, as this is likely to be an enhancement to the accommodation when re-letting the office space in the open market.

As with user covenants, the Landlord and Tenant Act 1988 does not apply to applications for licences to permit alterations to the demised premises. Thus, if the tenant's covenant against alterations contains a statement or proviso that consent shall not be unreasonably withheld, the landlord is not under a positive duty to give consent within a reasonable time nor to show that any condition or refusal of consent was reasonable.

Compensation for tenant's improvements

This is a much neglected right for tenants to claim compensation for improvements and the provisions are found under Part I of the Landlord and Tenant Act 1927.

A tenant will be entitled to compensation for improvements under Part I of the Landlord and Tenant Act 1927 if:

(a) before making the improvements, it applied to the landlord for consent
(b) the tenant leaves at the termination of its lease (howsoever determined)
(c) there has been a "qualifying improvement" to the premises
(d) the tenant claims in the prescribed manner
(e) the improvement(s) were carried out during the current tenancy.

The improvement must add to the letting value of the tenancy at its termination and must not be made in pursuance of a contractual obligation under the lease. Although it is not possible to contract out of Part I of the Landlord and Tenant Act 1927, the ability for tenants to claim this type of compensation, in our experience is very rarely used, largely out of ignorance, but also due to the rigorous procedure which must be followed by the tenant in following the steps which have to be taken, which might delay the matter at a time when the tenant will probably be more concerned in actually having the improvements carried out.

The claim must be made for compensation within three months of either the notice to quit, or the section 25 notice or re-entry by the landlord. The amount of compensation can not exceed either the net increase in the value of the holding made by the improvement or the reasonable cost of carrying out the improvement at the termination of the tenancy less the cost (if any) of putting the improvement in a reasonable state of repair.

If the landlord is proposing to make structural alterations, demolish or change the use of the holding, this may well have the effect of negating the net increase in the value of the holding caused by the tenant's improvement and in this instance there is unlikely to be any quantifiable compensation.

Disregard of tenant's improvements upon review

This is usually achieved in the rent review clause by either actually disregarding the tenant's improvements themselves, or disregarding the effect on rent of the tenant's improvements. This can create difficult valuation issues between the parties, and the mere fact that the tenant three years prior to the rent review date, carried out £100,000 of work relating to tenant's improvements, does not in itself mean that there will be a substantial reduction in rent due to the tenant's improvements, which may have been particularly relevant to the tenant's style of business and not actually improvements which enhanced the intrinsic rental value of the accommodation. A more recognisable tenant's improvement is where the tenant adds a facility to the accommodation which did not exist previously, such as the installation of a metal tile suspended ceiling, where previously the specification of the accommodation was purely a plastered and painted ceiling. In this type of situation, the valuer will try to find comparable evidence of properties with metal tile suspended ceilings and those with purely plastered and painted ceilings, so as to ascertain the difference in rental value between the two. This is not always the easiest proposition and an alternative approach is to amortise the cost of the tenant's improvement over the term of lease to be valued at rent review and deduct this from the rental value of the improved accommodation.

The rent review clause will normally allow the landlord to take into account on review tenant's improvements that were either:

(a) carried out in pursuance of an obligation to the landlord or
(b) carried out more than 21 years before the review date.

If there is not a disregard of tenant's improvements within the rent review clause, the premises will be valued as seen and this can place the tenant in a very disadvantageous position, in that not only is it having to pay for its improvements during the lease, but also potentially having to pay an enhanced rental value, reflecting the improved nature of the accommodation. There is a circular argument which can be employed here in the hypothetical letting, as the well advised tenant will argue that any tenant in the market, acquiring a lease without a disregard of tenant's improvements, will be placed in a disadvantaged position, when compared to tenants taking leases with the usual disregard of tenant's improvements, and for this reason will argue a discount from the market rent perhaps of 5% to reflect the onerous nature of this type of lease. It is rare today in a properly drafted institutional lease, for tenant's voluntary improvements not to be disregarded from consideration in the rent review hypothesis.

The Parties' Contract

At common law, until 1 January 1996, the concept of "privity of contract" bound the original landlord and the original tenant to the terms of their lease for its term. It was still the rule as institutional leases were being shaped and, as we have seen, was one of the attractions to an institution of property investment.

As the property boom of the 1980s gave way to recession in the early 1990s some pigeons came home to roost. That some of these pigeons were more verminous than their predecessors was not just a consequence of the privity of contract regime; it was also a function of the post-war economy and changes in the way the property market had chosen to deal with those changes. Inflation in the late 1960s and 1970s had led to the introduction of rent review provisions into rack rent leases — an attempt to compensate the landlord for the anticipated decline in the value of money during the life of the lease. These rent reviews were invariably upwards-only. The tenant who signed a 25 year lease with upwards-only rent reviews in 1979 may have assigned it in 1988 and retired (or, if it was a large multiple retailer, forgotten about it). No doubt the outgoing tenant would have taken an indemnity from its assignee at the time. But if the assignee subsequently defaulted and became insolvent, the landlord would cast around for someone to pay the rent. The original tenant would remain liable and, of course, there would have been a rent review in what was a very hot market in 1989. Imagine the tenant's horror as it received a rent demand for premises it no longer occupied, at a rent far in excess of that which was current when it assigned its lease, with no indemnity from its now-insolvent assignee and with the prospect of receiving demands to pay at least as much rent until 2004!

The political unacceptability of seeing retired people treated in this way (and not a little pressure from the large multiple-occupier organisations such as the British Retail Consortium) led to the passing of the Landlord and Tenant (Covenants) Act 1995. There was much controversy about this and the opponents to the new legislation (mainly landlords) saw this as something of a "knee-jerk" reaction to the property recession of the early to mid 1990s. This Act affected many aspects of the institutional lease, as we shall see.

A related area, and one which came under the microscope after the passing of the 1995 Act, is the alienation provision in the institutional lease. We have already seen that, before 1996, the original tenant remained liable. (So, in practice, did intermediate tenants because institutional leases would typically require assignees to give direct covenants to the landlord replicating the original tenant's liability.) Subject to the terms of its lease, a tenant had the right at common law to assign or sublet the whole or part of the demised premises. Indeed, the common law would not imply that a covenant not to assign or underlet without the landlord's consent was a "common or usual" one. Did this mean that an institutional landlord, having secured a respectable and financially strong first tenant, would reflect the thinking of the common law and give the first tenant some latitude in how it dealt with its lease? No. The institutional approach is to secure control and then (perhaps) make concessions. In some situations this control is quite reasonable, eg to maintain tenant mix in a shopping centre. But, in a single-let office block or solus retail unit, it is hard to see why an institutional landlord should have felt it necessary to seek such a degree of control — at least, before 1996.

The Landlord and Tenant Act 1927 ("the 1927 Act") had already qualified any consent required in the cases of:

- assignment
- underletting
- parting with possession
- charging.

It does not apply to absolute covenants or to the absolute parts of an otherwise qualified covenant (eg to procure a guarantor on an assignment).

In the 1960s and 1970s landlords contrived another way of

controlling dealings which was held not to fall foul of the 1927 Act. This was to require the tenant to offer to surrender its lease to the landlord before assigning or underletting. Where the lease was of business premises so that Part II of the Landlord and Tenant Act 1954 applied, a curious result occurred. While the tenant could be obliged to offer to surrender, if the landlord accepted that offer the consequent agreement was unenforceable and void. Some landlords may have felt that this was nevertheless a good way of preventing assignment. But it would also be a source of argument about the rental value of the premises on rent review, as we shall see in Chapter 5.

By 1980 the alienation clause in an institutional lease would usually contain the following restrictions:

- assignment or charging of whole permitted with qualified consent
- assignment or charging of part prohibited
- underletting of whole permitted with qualified consent
- underletting of part sometimes permitted with qualified consent.

Justice delayed is justice denied. Some landlords would delay responding to the tenant's applications for consent to assign or underlet in the hope that the "problem" of having the identity of the tenant or occupier of their premises change would go away. The tenant would have to seek consent first by stating accurately what was proposed and giving the landlord a reasonable time to reply. In difficult markets, or where the prospective assignee or undertenant required access soon (eg to fit out in time for a crucial trading period) the tenant would want a rapid response. For the landlord, dealing with the application was just another piece of administration. Agreeing to it would lead to a change which it would instinctively prefer to avoid. Why hurry? If things went on too long the tenant's options were to assign without consent (risky) or seek a declaration that consent was being unreasonably withheld (time consuming). The prospective assignee might lose interest in either case.

In an effort to redress the balance, the Landlord and Tenant Act 1988 ("the 1988 Act") was introduced. It applies where the tenant covenants not to enter into any of the following transactions without the landlord's qualified consent:

- assignment
- underletting
- parting with possession
- charging.

The 1988 Act does not apply where there is an absolute covenant against any of these activities. Nor does it affect the concept of reasonableness which has been built up over time by case law. But where landlord's consent is required to one of the transactions listed above and that consent is not to be unreasonably withheld, it will apply.

The 1988 Act creates a statutory duty for the landlord to consent within a reasonable time except where it is reasonable to refuse consent and, in either case, to give written notice to the tenant of the decision including details of conditions where consent is given conditionally and reasons for withholding consent where it is withheld. Thus an application delayed on an unreasonable basis (eg tenant's refusal to give an open-ended commitment to pay the landlord's costs of dealing with the application) or dealt with in a reasonable time but subject to an unreasonable condition will breach the landlord's duty. Failure to communicate its reasons in writing within a reasonable time will preclude the landlord from relying on them to justify refusal later. The key points are:

- it is only reasonable for a landlord to withhold consent to transactions which, if completed by the tenant, would amount to a breach of covenant
- the 1988 Act reverses the burden of proof as to reasonableness (formerly it was for the tenant to demonstrate that the landlord was acting unreasonably; now the landlord must show that he acted in a reasonable way and in a reasonable time)
- the remedy against the landlord is a personal one to the tenant which does not affect previous or subsequent owners of the reversion and is for breach of statutory duty.

While the 1988 Act does not define "reasonable time" which is a matter of fact in each case, a view has evolved that the industry standard is 28 days. It runs from the making of the tenant's application but if the information supporting the application is not supplied promptly, the time for responding is correspondingly extended.

As we have seen, institutional landlords like control. This control sometimes appears almost to be an end in itself. No doubt there are reasons why a superior landlord would wish to oblige its tenant to obtain consent to the assignment of an underlease by the undertenant in circumstances where there is a possibility that the undertenant may become the superior landlord's direct tenant at the end of the headlease (a possibility which is often removed elsewhere in the alienation clause by requiring any underlease to be outside the security of tenure provisions of the Landlord and Tenant Act 1954). Anyway, the practice was sufficiently widespread by the late 1980s for the 1988 Act to deal with the situation. In effect the superior landlord is subject to all the same duties as the immediate landlord, except that it must serve its responses on both its tenant and the undertenant. More bureaucracy for the institutional landlord but bureaucracy that must be handled efficiently to avoid the tortious claim of breach of statutory duty. This is a civil claim which may be for damages or an injunction, or both. The tenant must prove its loss.

The Landlord and Tenant (Covenants) Act 1995 ("the 1995 Act")

The unstoppable force of the institutional lease with its bond-like qualities met the immovable object of the 1995 Act with effect from 1 January 1996. In this section, we consider the effect on pre-1996 leases and the current environment.

The 1995 Act did not change the rules on privity of contract for pre-1996 leases. In a repeat of the soothsayers' reaction to the introduction of VAT as a potential addition to rent, there was much excitement about a two-tier market, with post-1995 leases commanding higher rents because of the release mechanism imposed by the 1995 Act (see below). But, as with VAT, this proved not to be the case. The alternative analysis, that investment property let on pre-1996 leases would benefit from improved investment values because of their more widely secured income stream, also turned out to be groundless.

For both pre-1996 and post-1995 leases, section 17 of the 1995 Act introduced the requirement for notice to be given in advance of making financial demands on former tenants. The demand must relate to one of the following:

- rent
- service charge (services, repairs, maintenance, insurance or the landlord's costs of management)
- liquidated sums made payable by the lease for failure to comply with a tenant covenant (this should be contrasted with debts — eg the recovery by the landlord of the cost of repairs carried out after default by the tenant which was held in *Jervis* v *Harris* to be a debt).

The section 17 notice must be served within six months of the date when the sum claimed fell due. If it is not, no claim can arise against someone who would otherwise be liable. For pre-1996 leases, potential recipients are:

- the original tenant
- other former tenants who gave a direct covenant to the landlord
- guarantors of a former tenant.

(The range of potential recipients of a section 17 notice is different for a post-1995 lease — see below.)

Renewal of pre-1996 leases

As we shall see, post-1995 alienation provisions are somewhat different from pre-1996 ones. The 1995 Act amended section 35 of the 1954 Act to make the court take account of the 1995 Act on renewal, so the landlord is not bound to offer a replica of the old lease provisions. However, in framing a proposal for an alienation provision in a renewal lease in the context of the changed law, the landlord is unwise to overplay his hand for two reasons. First, the imposition of provisions which are more onerous than current market practice may serve to reduce the rent that the tenant is willing to pay, either at renewal or at rent review. Second, such limited litigation as has been generated by the inability of the landlord and tenant to agree an alienation provision on a renewal of a pre-1996 lease suggests that the court will not support a landlord who seeks to justify onerous assignment provisions — in one decided case, the requirement for the tenant to provide an authorised guarantee agreement in any circumstances where the lease was assigned — even though the landlord argued that it was reasonable in the light of the changed privity regime. It is clear that the courts will not operate the 1995 Act to provide privity in another guise. One of the stated purposes of the 1995 Act was to "make provision for persons bound by covenants of a tenancy to be released from such covenants on the assignment of the tenancy". From this, it is clear that the rules of the game have fundamentally changed.

Post-1995 leases

We have seen how statute encroached on the landlord's ability to control dealings with its property under the Landlord and Tenant Acts of 1927 and 1988.

To compensate the landlord for the loss of the advantage of privity, the 1995 Act modifies the 1927 Act by allowing the landlord and the tenant to agree any terms for the giving or withholding of consent on an application for licence to assign by the tenant. Note that this change does not apply to residential or agricultural tenancies which are outside the scope of this book anyway. Neither does it apply to covenants against underletting or charging.

The landlord and the tenant may agree:

- circumstances in which the landlord may withhold consent to assign and
- conditions to which consent to assign may be subject.

This agreement must be entered before any application to assign to which it relates, to avoid the landlord introducing terms designed to avoid giving consent to a specific application. Nevertheless the change does represent an opportunity for the landlord to devise a strategy for preventing the dilution of the tenant's covenant strength. The strategy must, however, be based on terms which comply with the 1995 Act. A term based on fact (eg whether the prospective assignee has produced a certain level of pre-tax profits for a specified period before the date of the application) is acceptable without more. A term based on the landlord's discretion or opinion (eg the prospective assignee must be someone who in the landlord's opinion has some specific attribute) is circumscribed by the 1995 Act. Such a term will be read subject to the requirement that the landlord forms his opinion reasonably or that his determination can be conclusively reviewed by an independent third party identified in the agreement. While it is not explicitly stated in the 1995 Act that the reasonable application of the landlord's discretion is to be determined by the considerable body of case law under the "consent not to be unreasonably withheld" formula, it is most likely that this would be the analysis that would be applied. The second alternative militates in favour of the introduction of an expert determination provision in the lease for this purpose. If such a provision is not included and the landlord does not agree to act reasonably then the 1995 Act disapplies the test.

When the 1995 Act was introduced, landlords' advisers expended much ingenuity in devising tests to apply to prospective assignees. Some leases containing these ingenious but sometimes impractical provisions survive and it will be interesting to see how the landlords fare on rent review (dealt with in Chapter 5). A working party of the Association of British Insurers, representing a range of institutional landlords, produced a report containing some draft clauses designed to provide a framework for controls on assignment. These included:

- assignments to associated companies
- financial standing grounds
- provision of an authorised guarantee agreement (see below)
- rent deposit
- superior landlord's consent
- adverse effect on the value of the landlord's interest in the premises.

We look at the revamped Code of Practice for Commercial Leases in Chapter 10 of this book. The original programme gave rise to a survey which revealed that, in leases granted between 1996 and 1998, over three-quarters contained tests as follows:

- provision of an authorised guarantee agreement
- no breach of covenant
- no rent or other payments due to the landlord to be outstanding.

Further, half contained a profit or net asset test or a requirement that the prospective assignee be as strong financially as the tenant. In 2000 the British Property Federation ("BPF") and the British Council for Offices ("BCO") jointly published, in the light of experience to date, a number of model clauses which reflected market practice at the time while attempting to balance the legitimate expectations of the landlord and the tenant. Interestingly, they rejected "fact" tests in favour of "condition" tests. Those suggested were:

- a rent deposit or bank guarantee where reasonable
- provision of an authorised guarantee agreement
- provision for any contractual guarantor to stand behind the authorised guarantee agreement (this raises an interesting technical point which provoked some academic comment when the 1995 Act was passed — see below)
- payment of outstanding rent.

Otherwise the BPF/BCO proposal was content to rely on the "consent not to be unreasonably withheld" formula. And, as time passed and further evidence of market practice became available, these were still the recommendations in the BPF/BCO model clauses published in February 2003.

Authorised Guarantee Agreements ("AGAs")

The 1995 Act provides for the release of tenants and guarantors on an authorised assignment of a post-1995 lease. But, as one door closes, another opens, because (depending on the circumstances of the assignment) the landlord may yet require the tenant — and, perhaps, its guarantor — to guarantee its assignee's obligations. While the 1995 Act has rendered direct covenants from assignees unnecessary and would not allow the extension of the tenant's or its assignee's liability beyond its assignment of the lease, the 1995 Act allows the landlord to require the outgoing tenant to guarantee the assignee's obligations until the assignee itself assigns the lease.

What is an AGA?

The 1995 Act requires:

- that the outgoing tenant agrees to guarantee its assignee's obligations in relation to covenants from which the outgoing tenant is released on assignment
- that the lease provides for the landlord's consent to assignment subject to a lawful condition that the tenant gives an AGA. The lawfulness of the condition will depend either on reasonableness under the 1927 Act (see above) or the specific introduction under the 1995 Act of a condition requiring an AGA to be given on any assignment and
- that the AGA conforms with the 1995 Act's provisions as to duration and content.

As to duration, the AGA may only require the guarantee of performance of the assignee of the covenants from which the tenant has been released for the period of that assignee's liability. Over-zealous drafting which seeks to extend liability beyond the assignee's or to include other matters is not fatal; the additional provisions are unenforceable but the basic guarantee is not.

As to content, the 1995 Act does not prescribe a form of AGA. Again, excessive enthusiasm in drafting will result only in the excessive terms being severed from the compliant part of the AGA. Provisions commonly found in AGAs which might run this risk include a requirement to take up a new lease if the original lease is forfeited and to repay up to six months' rent if the landlord decides not to require the guarantor to take up a new lease. However, disclaimer is well established as releasing only the current tenant. The current tenant's guarantors and those remaining liable whether under AGAs or as guarantors at common law under a pre-1996 lease remain liable. The 1995 Act permits the landlord to include an obligation to take a new lease for the remainder of the term of the disclaimed lease and from a management viewpoint the landlord would be advised to do so.

Excluded assignments

Not every assignment brings the possibility of an AGA into play or releases the assigning tenant from its lease obligations. The 1995 Act provides that assignments by operation of the law (eg to a tenant's trustee in bankruptcy) or in breach of covenant (eg where the tenant fails to obtain the requisite consent) are "excluded assignments". The original tenant remains liable on a joint and several basis with its assignee until a non-excluded assignment takes place. At that point both may be required to enter an AGA if circumstances dictate (see above). The AGA providers will not then be released until another non-excluded assignment takes place.

Guarantors and AGAs

The common thread of the institutional landlord's approach is to maximise the number of potential payers of rent and performers of obligations where a lease is assigned. The 1995 Act was not, unfortunately, entirely clear about the role of contractual guarantors and AGAs. Suppose the first tenant (T) is guaranteed by its parent company (G). T wishes to assign. The lease provides for an AGA. Is it enough for T alone to give the AGA, or can the landlord require G to guarantee T's obligations under it? This was a matter of some academic debate just after the 1995 Act was passed. The 1995 Act provides for G to be released to the same extent as T when T was. Further, G cannot be compelled to give the AGA; it can only be given by T and, by definition, G is not T. Yet the view has developed that the landlord is in luck — G can be required to guarantee T's obligations under the AGA. The argument revolves around the distinction between a "tenant covenant" from which T is entitled to be released on assignment and the AGA itself which is *not* a tenant covenant. There is no reason in principle why G should not continue to stand behind T in the AGA (although G cannot be required to enter the AGA as a guarantor of the assignee — only T can do this). The landlord therefore drafts G's obligations to include a guarantee of T's obligations under an AGA. This is, the landlord will argue, a reasonable commercial approach. It does not, the landlord will claim, amount to privity in another guise. It simply maintains the covenant strength of the occupying tenant in a way anticipated by the 1995 Act. The reader is reminded that there is no decided case on this point yet and would be forgiven for wondering — if the analysis set out in this paragraph is followed — what all the fuss was about.

Section 17 notices

As we have seen, these can be served in relation both to pre-1996 and post-1995 leases. The difference in the post-1995 leases is that the range of potential recipients is limited to:

* a former tenant with subsisting liability under an AGA
* a former tenant who made an excluded assignment and
* a guarantor of either who has not been released.

Section 17 notices offer the landlord an opportunity to angle for a new tenant. As we shall see, a former tenant who pays the full amount due in response to a section 17 notice is entitled to a new overriding lease. The landlord may — indeed, should — go "tenant shopping" and, assuming not more than six months' arrears have accrued, pursue former tenants in order of covenant strength, starting with the strongest.

The form and content of a section 17 notice is dealt with in the 1995 Act. It must list the sums claimed, describe them (eg rent, service charge) and include the basis of interest if that is claimed too. The landlord may not claim a sum in excess of the amount specified (excluding interest). Where figures are provisional a further notice must be sent within three months of the figure becoming firm. Perhaps curiously, the prescribed form of notice does not alert the recipient to its right to claim an overriding lease on payment of all amounts claimed.

Overriding leases

Once a former tenant or guarantor has received a section 17 notice and paid the amounts claimed it has a period of 12 months from the date of payment to claim an overriding lease. The landlord must grant it within a reasonable time of request and the claimant must pay the landlord's reasonable costs. The claimant has no rights under the overriding lease until it has given the executed counterpart to the landlord. This is crucial as the only reason for the claimant to take the lease is to deal direct with the defaulting tenant (eg by forfeiting that tenant's lease).

The 1995 Act contains basic rules for the term and content of the overriding lease. The term will usually be three days longer than the existing lease. Where the landlord does not itself have a reversion to the existing lease exceeding three days (unlikely in an institutional situation) then a shorter reversion will have to be granted. The obligations in the overriding lease will reflect those in the existing lease with the omission of personal or completed obligations. The rent review should be to the same effect as in the existing lease — it could be disastrous for the claimant to find itself paying a higher rent than the defaulting tenant as a result of inconsistencies, for example because the disregard of tenant's improvements were not extended to those of undertenants (which the defaulting tenant will be for the purposes of the overriding lease) or an assumed term for the hypothetical lease at review reflecting the original term of the lease under review (where the terms will by definition be different and in consequence could lead to different rental values).

While the 1995 Act deems consent from mortgagees to the overriding lease, it does not deem consent from superior landlords. The best view seems to be that it would be unreasonable for a superior landlord to refuse consent to a mesne landlord complying with a statutory duty; but an application should still be made. If the superior landlord is difficult for whatever reason, the mesne landlord may find itself defending a claim for breach of covenant.

So far, so good? Perhaps (subject to the points made in the previous paragraph) for the landlord. Legislation designed to help tenants against the iniquity (in some situations) of privity of contract has created different problems for the claimant. The range of possible actions for a prospective claimant is bewildering, even with legal advice. One certainty is that it will have to pay Stamp Duty Land Tax on the overriding lease. Another is that it will be managing a property containing a recalcitrant occupier. It seems likely (but is not entirely clear) that the occupier must default again before the claimant can forfeit the original lease. And, if the original lease contains a turnover rent, the claimant's lease will have to reserve a rent equivalent to that turnover rent. One hopes for the claimant's sake that turnover is being verified by electronic point of sale information being transmitted directly to the superior landlord as is often the case with shopping centres, or its task of policing and securing payment from the occupier will be extremely burdensome. (In practice this may not prove to be a problem. Most turnover centres are extremely actively managed by their owners who would probably see more benefit in removing a failing tenant than enabling a third party to claim an overriding lease.) And what of overrented premises? Perhaps the real benefit of the 1995 Act for tenants is not its specific provisions so much as its contribution to an environment where shorter leases are more attractive — a recurring theme in our exploration of the institutional lease.

A sting in the tail for the landlord

As we have seen, the 1995 Act abolished (or at least drastically modified) liability under privity of contract for tenants. But what of the landlord's liabilities?

Pre-1996

The original landlord has always been liable on its covenants for the duration of the lease. Subsequent transferees of the reversion were liable while they held it. A tenant does not usually have any control over the identity of its landlord — in contrast to the extensive controls usually available to the landlord when the tenant seeks to assign its lease. This could lead to a situation, in theory at least, where the landlord might be a man of straw and the tenant would have no effective rights against its predecessors. This theory overlooks the commercial reality that the property itself which was the subject of the tenant's lease would inevitably have some value.

Post-1995

The 1995 Act made any landlord liable on a continuing basis for its obligations under a lease unless it obtained a release of those obligations before or within four weeks after the transfer of its reversion. Thus the landlord is arguably in a worse position than the tenant since 1995 as, not only is it not automatically released from its covenants, but a liability which did not exist before 1996 was created by the 1995 Act. Apparently the Law Commission's analysis of the landlord's position was that, as it does not usually have many obligations (covenants for quiet enjoyment, to insure and where appropriate to provide services) it would not matter if its potential liability to perform these obligations were extended. One might, of course, argue precisely the opposite — because there are few obligations, from what mischief is the tenant being protected? But that analysis is misplaced. For example, the insurance obligation is critical and, if not observed, potentially disastrous in terms of cost. An indemnity from the transferee of the reversion may or may not suffice dependent on the financial standing of the transferee. However, that indemnity itself may be problematic — see below.

Mechanics for release of an outgoing landlord requires it to serve a notice on the tenant during the period referred to above. This process may not be straightforward. If the tenant agrees to the release or does not respond within four weeks of the notice, the landlord is released at the price of losing the benefit of the releasing tenant's covenants under the lease. If the tenant objects, the landlord can apply to the county court for a declaration that it should be released. No statutory guidance about grounds for objection appears in the 1995 Act but one would assume in the absence of litigation on the point that the proposed transferee's financial ability to perform the landlord's lease obligations will be a crucial element in deciding any such proceedings.

Multiple occupiers may not all act consistently when invited to release the outgoing landlord. The 1995 Act does not require unanimous acquiescence or objection by the tenants of a multi-let building, nor does it enable the majority to bind the whole. If some object and there is a subsequent problem which requires expenditure from the outgoing landlord, its redress will be limited to those who objected for, as we have seen, the release operates both ways. At least one commentator cites this problem as an argument against the landlord seeking a release where the premises are multi-let and there is doubt about whether that release will be forthcoming from all the tenants.

A landlord who is not released on his own disposal has a second bite of the cherry on a subsequent disposal by its transferee. Of course, to take this bite, it must know that the transferee is itself disposing of the property and it would increase its chances of success if it could point to a prospective transfer to a reputable and financially strong body. Covenants to provide this information are increasingly sought on disposals where the issue of release is a live one.

The indemnity problem — Conveyancers routinely seek indemnities from transferees in relation to continuing liabilities such as those that are faced by an unreleased landlord. Under the 1995 Act, such an indemnity might be avoided to the extent that it sought to impose continuing liability on the transferee of the reversion beyond the date when that transferee is itself released (albeit by a different party, the tenant) on its own disposal of the reversion. The most satisfactory technical solution to this problem appears to be to require an indemnity from the transferee for as long as it owns the reversion coupled with a covenant by the transferee to obtain a corresponding indemnity and covenant from its successor for so long as the original landlord is unreleased.

Variations: impact on former tenants and guarantors

Having read this far, the reader may have formed an impression that the institutional lease is not designed to provide flexibility or latitude for the tenant. Rather is it designed to create investment value for the landlord. When reporting on the content of such a lease — a task which befalls the average commercial property lawyer many times a year, whether the client is a landlord, tenant or lender — a key area is what the tenant may (or, more usually, may not) do. In the areas where the tenant is committed (eg rent payment) or, for whatever reason, has no scope for change (eg a specified user clause with no provision for change) the client and its adviser may reasonably assume that that, so to speak, is that. The lease is conclusive; and, of course, it is, but it is never immutable. However prescriptive or restrictive the terms of the document, if the parties subsequently agree to change it, they can. That decision may be provoked by changing market circumstances. Often the tenant will offer more rent if a restriction is removed. In a recession, the landlord may make a concession about rent rather than lose the tenant through insolvency.

Yet, as we saw earlier in this chapter, the name of the game for the institutional landlord is to secure a rental stream for as long as possible with recourse to as many current and former tenants and guarantors as possible to pay it. The 1995 Act introduced rules to release former tenants from liability on assignment. But even before 1996 it was possible for the landlord — usually inadvertently — to release a former tenant and its guarantor, and even the guarantor of the current tenant, by varying the terms of the lease. As we shall see, since 1995 section 18 of the 1995 Act has attempted to introduce consistency to the effects of various kinds of variation.

Implied surrender and regrant

It is possible for a variation to take effect as a surrender and regrant. While this can be highly disadvantageous to the landlord — it has the effect of releasing everyone except the current tenant — it is not always anticipated by surveyors or solicitors. Experience shows that this outcome has been inadvertently achieved on occasion. The concept survived the 1995 Act. The test of implied surrender and regrant is — can the variation be effected without granting a new lease? It will be triggered by any of the following variations to the terms of the original lease:

- extending the term of the lease (safe alternative — grant a reversionary lease)
- adding to the premises demised by the lease (safe alternative — grant a supplemental lease of the additional premises)
- introducing exclusion from the protection of the 1954 Act
- introducing a landlord's break right in conjunction with 1954 Act exclusion.

It can also be triggered by either of the following unless the variation is documented correctly:

- reducing the term of the lease
- reducing the extent of the demised premises.

Old leases

For pre-1996 leases, the effects are potentially very far reaching. All previous tenants and guarantors and, indeed, the guarantor of the current tenant will be released if the variation is entered between the current landlord and the current tenant. If the current guarantor is to be bound, it must be a party to the variation.

New leases

Here the effect of the surrender and regrant is potentially less damaging, because the 1995 Act will have limited the liability of former tenants and their guarantors. Nevertheless there will be no further liability for any of them under the surrendered lease. The landlord will also "lose" any guarantor of the current tenant unless it is made a party to the variation.

All cases

The parties to the variation effecting the surrender and regrant will be the parties to the new lease. If the surrender and regrant takes place after 1995 the resulting tenancy will be a new lease for the purposes of the 1995 Act, with all the consequences for original tenant liability that that brings (see above). Further, because it is the grant of a new lease, it will have to be stamped and (if for more than seven years) registered at the Land Registry. It may even cause a problem on the tenant's balance sheet depending on the rules prevailing at the time (see Chapter 9 for a consideration of all these topics).

Variations not amounting to a surrender and regrant — before the 1995 Act

The tenant's problem

Until 1996 previous tenants (either the original tenant or any assignee who had given a direct covenant before itself assigning) ran the risk that the lease terms might be changed by the tenant for the time being. This would result in a potential extension of the original/earlier tenant's liability despite the fact that it was powerless to prevent the variation, could not object to it and did not need to be a party to it. The classic example is a variation to the user clause. A restrictive user provision leads to a lower rent because the category of tenants willing to bid for the premises is smaller. If the user clause is widened to admit more prospective tenants, or changed to admit a particular class of user that is willing to pay a higher rent for whatever reason (often scarcity or location), the rent will very likely be increased immediately to reflect that variation. Even if it is not — the current tenant might pay a capital sum to secure the change — any subsequent rent review will have to reflect the change assuming that the hypothetical lease is on the terms of the actual lease as varied (see Chapter 5). This rental revaluation would bind the original/earlier tenants if the current tenant defaulted. Not surprisingly, the original/earlier tenants would try to reduce their liability in such circumstances. Although a total release was unlikely to be forthcoming through the courts, the increased liability engendered by the variation was seen as fair game. Unfortunately for the original/earlier tenants, for many years the court's analysis was short and sweet. An 1888 case in the Court of Appeal was authority for the proposition that the assignor of a lease gave the assignee all the powers which the assignor had in relation to the lease. These included the ability to vary it. Thus the assignee could agree a variation coupled with or leading to a significant increase in the rent and then default, leaving the original/earlier tenants to meet their new, unexpected liabilities.

A stitch in time?

It was not until the mid 1990s that this clear injustice was redressed. In 1995 the Court of Appeal had the opportunity of analysing the 1888 case and pronounced that the earlier interpretation was wrong insofar as it related to the impact of lease variations on the original/earlier tenants. The essence of the new analysis was that, where a variation is anticipated by the terms of the lease (eg the review of rent), the original tenant would be bound to comply with the varied obligation even though the change took place after the assignment. If, however, the variation were outside the terms of the lease — such as the change of use example set out above — while the variation would be valid it would not affect the original/earlier tenants' obligations. This judgment itself anticipated section 18 of the 1995 Act in many respects, although some issues remained outstanding:

- Many leases contain a definition of "the lease" which includes subsequent deeds of variation. A former tenant may struggle to convince the court that a subsequent variation, even an extra-contractual one, does not fall within this definition. (There was no such definition in the lease considered in the 1995 case.)
- The effect of the variation on rent at review is often contemplated by the rent review clause. The hypothetical lease which is to be valued at review is often — although, curiously, not in the Law Society/RICS Model Form — defined to include variations to the form of lease originally granted. It is a moot point whether this would include extra-contractual variations to the lease or is merely a way of anticipating changes to the original lease terms (eg the fact that the rent will be reviewed). It certainly appears from a subsequent case that the original/earlier tenants and their guarantors will be saved where the definition of the hypothetical lease refers to the form of the lease as granted without regard to subsequent variations.

The guarantor's position

If the variation prejudices the guarantor, the guarantor will be discharged unless:

- the guarantor has consented to the variation or
- the variation does not prejudice the guarantor or
- the drafting of the guarantee means that the guarantor's obligation survives the variation (note that the courts will interpret any such provision against the landlord and very restrictively).

A variation not falling into any of these exceptions and made by the original tenant without the participation of the guarantor will release the guarantor. However, after the 1995 case referred to above, a variation by an assignee which is outside the mechanisms contained in the lease will not affect the liability of the original tenant. A guarantor of the original tenant will hence not be discharged because it is not prejudiced by the variation.

Variations not amounting to a surrender and regrant — after the 1995 Act

The relevant law is contained in section 18 of the 1995 Act (in this chapter, "section 18") which applies to both "old" and "new" leases.

The purpose of section 18 was to give statutory protection to original/earlier tenants from the effects of a "relevant variation". A relevant variation is one which does or might impose on an original/earlier tenant liability to pay an increased sum and to which the landlord was not obliged to consent when the original/earlier tenant assigned the lease. This may be because either:

- the landlord had an absolute right to reject the variation or
- the landlord agreed a variation subsequent to the assignment by the original/earlier tenant by virtue of which it has lost the absolute right to reject the variation which it had at the time of the earlier/original tenant's assignment.

The following variations are outside the scope of section 18:

- variations to which the landlord cannot unreasonably withhold consent (remember that consent may be qualified in this way either because of the precise wording of the lease or because statute has added them — for example, where consent is required for a dealing with the lease — Landlord and Tenant Act 1927)
- variations which reduce the tenant's liability
- variations creating liabilities which the original/earlier tenants undertook voluntarily when the variation was made
- variations affecting the landlord's liability.

Any variation must, of course, be carried out in accordance with the necessary formalities to work. Usually this means it should be effected by deed. Section 18 applies to all variations however effected.

Section 18 and new leases

At first blush the reader could be forgiven for thinking that section 18 would have little application to new leases. Unless there has been an excluded assignment, previous tenants will have been released on assignment by virtue of the 1995 Act. However, in practice many former tenants remain liable under the authorised guarantee agreement which they will have been reasonably required to enter on the assignment of the lease. So, former tenants remaining liable as tenants as a result of an excluded assignment now have the protection of section 18; former tenants entering authorised guarantee agreements are liable as guarantors, as to which, see below.

Section 18 and old leases

Here the universe of beneficiaries of the limitations contained in section 18 is potentially wider. All original/earlier tenants who gave direct covenants and their guarantors remain liable but have the benefit of section 18. As to the guarantors, their position is the same as the giver of the authorised guarantee agreement referred to in the previous paragraph.

The effect of section 18

For original/earlier tenants, the effect is to relieve them from liability to pay any sum due to the landlord which results from a relevant variation made after that tenant assigned the lease. Readers are referred to the example of the varied user clause earlier in this chapter. Now, on the basis of the following fact matrix, the position would be as follows:

- L grants lease to T in 1990 with single retail use restriction in user clause
- T assigns lease to A1 in 1993; A1 gives a direct covenant to L
- In 1994 L and A1 vary the lease to widen the user clause to permit one of three defined retail uses in consideration of a rent increase of 10%
- In 1995 there is a rent review
- In 1997 A1 assigns the lease to A2
- In 1999 L and A2 vary the lease to allow an open retail user in consideration of a rent increase of 15%. A1 is not a party to the variation
- In 2000 there is a rent review
- In 2002 A2 defaults on its obligations to pay rent.

T's liability

The 1994 variation is not a "relevant variation" as it pre-dates the 1995 Act. T will not have the benefit of section 18 in relation to it. However, the 1995 case referred to above was to substantially the same effect and T can rely on this to restrict T's liability to that which would reflect the original terms of the lease as granted in 1990. The 1999 variation is a relevant variation so, by virtue of section 18, T will not be liable to pay any part of the increase attributable to it. These limitations on T's liability may lead to evidential difficulties, as the 1995 and 2000 rent reviews will have been conducted on the basis of the lease as varied at the relevant date. It is highly unlikely that any thought will have been given to the correct rental to reflect T's original liability. It may become necessary to conduct another rent review to settle this point. The associated procedural difficulties are reviewed at the end of this chapter. Unfortunately, we cannot see a ready solution to them.

A1's liability

A1 was a party to the 1994 variation so will be liable for any rent or increase in rent at review attributable to it. It was not, however, a party to the 1999 variation so the same analysis as that for T applies. Section 18 would also operate to help A1 as the variation took place after 1995. Again, the evidential issue about the correct level of rent for which A1 is liable (this time, on the basis of a rent review in 2000 of a lease as varied by the 1994 variation) will arise.

The guarantor's position following a variation

The common law position as set out earlier in this chapter still applies, but now section 18 has introduced the concept of a relevant variation. It provides that, if a guarantor's liability is not wholly discharged by the variation of the tenant's covenants in the lease, its guarantee of a former tenant's liability is limited so as not to include any amount payable as a result of a relevant variation. The key words here are "former tenant" — the guarantor of the current tenant, whether or not by virtue of an authorised guarantee agreement, does not have the protection of section 18. Nevertheless, as we have seen, the common law is solicitous of a guarantor's rights and the safest course for any landlord contemplating a variation of a lease where the tenant is currently guaranteed is to join the guarantor as a party to the variation. Where the tenant and its guarantor are associated in some way (eg where the guarantee is given by a director of the tenant, or its parent company) this course may not be a problem. Where the guarantor is a former tenant who has given an authorised guarantee agreement, the landlord may well find that it cannot secure participation in the variation by that guarantor. In that case the landlord must weigh its options carefully — is the increase in rental value worth the loss of an additional party's obligation to pay it?

Some other practical problems

- While it is impossible to contract out of section 18, the 1995 case may or may not have application depending on the precise wording of the lease/variation being considered. This is a complicated and unusually technical area and the outcome can only be predicted on a case by case basis.
- Where the landlord wishes to serve a default notice on an earlier tenant and there has been a relevant variation which has increased the rent beyond that tenant's liability as the result of a subsequent rent review, the landlord must choose either to seek the amount for which it thinks the former tenant is liable (which may involve an argument with the former tenant before the amount due can be settled) or claim the full amount (which may lead to the former tenant arguing successfully that the notice is invalid). Neither course is attractive where, as we saw above, time is of the essence for the service of the default notice.

Conclusion

The reader will have seen that this is a difficult area, both commercially and legally, and has been for years. By the time institutional leases were being routinely granted for 25 years in an inflationary environment, the parties had to make provision for changes in the rent to reflect the inevitability of changes to the value of money during the lease term — a topic dealt with in the next chapter. But other changes take place to the economic and social environment in 25 years as well, and the generally prescriptive nature of the institutional landlord's contract with its tenant set out in the institutional lease was never likely to anticipate them. It is objectively reasonable for the landlord and the tenant to be able to change the terms of the lease contract to reflect their mutual desires or meet external pressures from time to time during the life of the lease. It was unfortunate for a former tenant when any such change might commit it to extensive additional liability. It was also unfortunate for the landlord if a small change to the contract had the effect of releasing earlier tenants and/or their guarantors from their commitments. The 1995 Act has performed a service to both by contriving both to limit former tenants' liabilities to those to which they have expressly agreed and, in doing so, entitling the landlord to the continuing benefit of guarantees given in respect of those liabilities. Yet, once again, it seems that the 1995 Act is a symptom of the problem rather than the cure. Variations of institutional leases become necessary because 25 years is a long time. Shorter leases offer the opportunity of renegotiation on a regular basis and obviate the need for fundamental alteration outside the mechanisms that a longer lease will usually contain, on a very limited range of terms.

Rent Reviews

Purpose

One can do little better than to quote Sir Nicholas Browne-Wilkinson V-C's observation in *British Gas Corp* v *Universities Superannuation Scheme* [1986] 1 EGLR 120:

> There is really no dispute that the general purpose of a provision for rent review is to enable the landlord to obtain from time to time the market rental which the premises would command if let on the same terms in the open market at the review dates. The purpose is to reflect the changes in the value of money and real increases in the value of the property during a long term.

We saw in Chapter 1 that institutions with "real" liabilities such as pension payment obligations prefer assets which are capable of maintaining value during periods of (sometimes, but not recently, high) inflation.

Development

The concept of rent review is over 150 years old. It seems that the idea of a variable rent was a commonplace by the late 19th century although the earliest reported case about a change in rent *during* the term was heard in 1926 (relating to the ten yearly revision of a ground rent under a lease of land in New Zealand granted in 1886). As is often the way with innovative leasing practices, there was some academic fluttering in the dovecotes about the legal basis for varying a rent by arbitration which eventually died down — but only just over 40 years ago. Indeed, the Law of Property Act 1969, which enabled the court to include a rent review provision in a renewal lease under the 1954 Act, was the first statutory acknowledgement of the use of rent review clauses.

Rent reviews grew in frequency in the late 1950s and 1960s generating much litigation in the 1970s. The late 1960s and 1970s were a period of high inflation and rent review intervals started to reduce in consequence. The 1950s had seen 42 year leases with 21 year review periods; in the 1960s these periods had reduced to 14, ten or even seven years and as late as 1965 the opinion was expressed that five years was the shortest interval that could reasonably be contemplated. The high watermark — so far — was reached when three years was occasionally used in the 1980s. Today, five years is the "industry standard" and this has generally been the case since institutional leases reached a developed state in the 1980s.

The machinery of rent review

A common feature of older (pre-1980) institutional leases was a provision for review by notice. These posed — usually unintentionally — risks for the landlord and, perhaps intentionally, risks for the tenant. It is probably correct to say that, by the mid-1980s, reviews by notice had largely been abandoned but there will still be some hanging on out there so some comment is in order.

- Who can initiate? Often only the landlord, although it is better if either party can initiate the procedure. (In rare non-institutional leases where the rent review can be up or down it is, of course, vital that the tenant should be able to initiate the review too.) If the tenant is unable to do this his ability to budget or assign will be severely circumscribed. The tenant may be able to serve notice on the landlord requiring him to initiate the review within a reasonable time but only if the lease does not reserve the right to review exclusively to the landlord and the review date has passed.

- "Desire" notice and evolution to "trigger" notice. As rent review procedures developed, clauses moved from "desire" notices (where the landlord simply told the tenant that it wished to review the rent without more) to "trigger" notices where the notice itself became part of the procedure because it specified the landlord's proposal for a new rent and, sometimes, started the clock running for a response by the tenant. The "desire" notice itself had fallen out of favour by the early 1980s (although curiously, even today, landlords often ask their solicitors to write to the tenants in terms of a "desire" notice even where the review clause requires no such step).

"Trigger" notices were prevalent at a time when the landlord was in a stronger bargaining position than is generally true today. The key issues were that the landlord (but not the tenant) could

- specify a new rent in the notice and
- start the time specified in the review clause running

In its most brutal form, the tenant would have to serve a counter-notice within the specified time (which was of the essence) otherwise the landlord's proposal would become the new rent. Inevitably such provisions led to disputes about the adequacy of the tenant's counternotice — disputes which could be very costly to the tenant if it lost.

Slightly less blatant were the clauses where the landlord's notice started time running for agreement of a new rent or, in default, reference to a third party (expert or arbitrator). The landlord's original proposal for a new rent would apply if agreement was not reached within the stated time, commonly three months. This might be a reasonable time for an expert determination. In the case of reference to an arbitrator in accordance with the Arbitration Acts, the arbitrator is able to extend the time-limit where undue hardship would arise from strict enforcement of the original timetable.

This gloomy tenant scenario was lightened only by two possibilities:

- the unlikely concept of the landlord being ignorant of the true market rent of the premises and proposing a low rent in the original notice which will bind him if the tenant accepts it or
- that time is of the essence for the service of the landlord's notice and it does not serve it within the specified time, thus losing the right to review.

The second possibility is not as outlandish as it may seem. In the 1970s and early 1980s, although there were precedents for rent review clauses, many were "bespoke" and none the better for it. The idea that making time of the essence worked to the landlord's advantage occasionally led to all provisions of the clause being subject to that rule, either through initial drafting error or cunning counter-amendment by the tenant's solicitors in negotiations. In our experience pre-1985 rent review provisions always require especially careful study. However, the well drafted rent review clause from the 1970s or 1980s are often a model of conciseness, producing a clear unambiguous directive to the valuers in less than a page, a far cry from some rent review clauses found in modern leases which sometimes extend to several pages and are unnecessarily verbose. Today, leases very infrequently contain time of the essence provisions in rent review clauses as both landlords and tenants have recognised the dangers that these entail. Although timing provisions are often used, these are normally advisory rather than mandatory such as the ability for either party if agreement of the rent has not been reached by the review date, to refer the matter to the President of the Royal Institution of Chartered Surveyors for the appointment/nomination of an arbitrator or independent expert, giving the review clause a structure. Similarly most rent review clauses now do not require a formal notice to be served by either landlord or tenant to activate the rent review. The review process will normally begin by the landlord simply contacting the tenant on or just before the actual review date, in order to inspect the premises and this will be followed by a quoting rent if appropriate and the ensuing negotiations. The arbitration or independent expert route will only be required if an agreement cannot be reached or if satisfactory progress is not being made and a stimulus is required to the negotiations.

A move towards standardisation

The first draft model form of rent review clause relating to commercial leases was published jointly by RICS and the Law Society in June 1979. This provoked sufficient comment and criticism from professionals and consumers to enable the first model form to be published in March 1980. A further edition was produced in 1985. The notes accompanying the 1980 edition reflected the issues it was intended to address. The model form probably dominated and at the very least heavily influenced the drafting of rent review clauses following its introduction:

- the clause was drafted to cover all kinds of business premises, from poor quality multi-lets to single-let modern buildings
- lease lengths might vary
- attempts to anticipate alternative market conditions with upwards only or upwards and downwards rent reviews, landlord-only initiation (see above) and choosing expert or arbitrator determination
- "time of the essence" was avoided to stop human error from vitiating the review
- no third party decision before the review date.

The clause itself, of course, sets out the procedure for the review and the basis on which the rental value of the premises is to be ascertained. Earlier in this chapter we explored some of the issues commonly arising from clauses drafted in the 1970s. The reader will have noted that they were predominantly "machinery" rather than "valuation" issues. After the arrival of the model form, it is generally true to say that the focus changed to valuation issues.

The use of model forms of rent review clauses and leases in general is considered helpful to the industry as a whole but it is the fact that every situation is different and the use of word processors has created a degree of carelessness, with the result being some inappropriate drafting eg the requirement for an RICS appointed independent expert to be a retail specialist when the subject property is a large air conditioned office building.

The rent review dates

It is desirable to specify the rent review dates rather than attempt to describe them by such referential methods as "the expiry of the fifth year of the term" or "each fifth anniversary of the term commencement date". These can create doubt, especially where the valuer is trying to decide the hypothetical lease term or where the outcome of a rent review is linked to the exercise of a tenant's option to break the lease (as to which, see Chapter 7). The use of descriptions linked to the expiry of years of the term or to anniversaries also raises the question of whether there is intended to be a review during any period of holding over or continuation of the tenancy by virtue of the Landlord and Tenant Act 1954. A review on the first day of any such continuation might be inferred if the term is defined to include any such period of holding over or continuation. If a review at the end of the contractual term is intended then it should be dealt with expressly as a penultimate or "last day" review. The purpose of such a review is to provide for a current market rent during any period of holding over or statutory continuation rather than relying on the provisions for interim rent contained in the 1954 Act or, if an interim rent application is not made, continuing with a rent set at least five years previously.

Valuation of the premises at rent review

The rent review clause should set out the basis of valuation of the premises. The RICS/Law Society Model Form deals with this by inviting the valuer to make a number of assumptions about the premises while directing the valuer to disregard other matters.

The essence of the assumptions is that they provide the factual matrix from which the valuer is to form a view. Some of those facts will reflect reality — eg that the premises are those demised by the lease — while others require the valuer to assume facts which are not necessarily reflected in reality — eg that the landlord and the tenant have observed and performed the lease covenants. A distinction needs to be made here in that it is unfair for an assumption that the landlord has complied with its covenants as opposed to a tenant. The landlord would benefit at review from a breach of its own covenants as compared to a tenant who does not comply, where the assumption should be made to ensure that the tenant did not benefit from its non-compliance. Where the courts are invited to pronounce in an absence of clarity in this drafting, they will usually favour a construction which reflects reality. But, as we will see, this is not always possible.

One assumption, which is made in the Model Form, is that the hypothetical lease is on the same terms as the lease under review. If the landlord has included a surrender-back requirement should the tenant wish to assign, such as we considered in Chapter 4, or forbidden underletting at less than the rent passing under the headlease, which can have the effect of preventing underletting altogether in a falling market, it is reasonable to assume that the hypothetical tenant will reduce its rental bid in the face of these restrictions.

Another assumption which gained popularity with landlords around 1990 when VAT was introduced to land transactions was that the hypothetical landlord and tenant would be able to recover VAT in full. This arose from a concern that an exempt or partially exempt tenant who could make either no or only partial recovery of VAT would put in a lower bid for rent than would otherwise be the case. The assumption was designed to protect landlords whose investments might be suitable for occupation by banks, financial institutions or other businesses with limited VAT recovery. Such assumptions are discouraged by the presumption of reality favoured by the court where possible (see below). If an investment was in fact occupied by an exempt or partially exempt business and/or was of a type that would be likely to attract such a business then it would amount to a direction to the valuer to depart from reality. As with headline rents, also considered below, this might result in a compensating discount. The preferred course now seems to be to ignore VAT in the rent review clause.

The disregards are designed primarily to avoid the tenant having to pay an increased rent for benefits which it has brought to the premises. The usual "disregards" are of occupation, goodwill and improvements, reflecting the provisions of section 34 of the Landlord and Tenant Act 1954. A disregard of tenant's improvements is essential if the tenant is to avoid paying twice for its improvements — once to make them, and again through a higher rent because of their utility in the premises. The improvements disregard was the subject of a recent negligence action which serves to highlight its potential impact on rents. The tenant had taken a site and developed it as a golf course including a clubhouse; the total cost of the development was £1.6m. The lease of the site included a covenant by the tenant to carry out these works. It was thus unarguable that they were carried out pursuant to an obligation to the landlord. The rent review clause included a disregard of tenant's improvements which, in the usual way, did not extend to improvements "made in pursuance of an obligation to the landlord". Consequently a valuer called upon to ascertain the market rent of the premises had to value it as a developed site as opposed to an undeveloped site with the benefit of planning consent. The difference was between an annual rent of £100,000 and one of £45,000. As the term of the lease was for 100 years with five yearly reviews, it is unsurprising to learn that the tenant alleged negligence on the part of its solicitors in accepting the rent review provision, with damages sought of the order of £750,000. In another case, the tenant successfully obtained rectification of its lease where the rent review clause failed to direct the valuer to disregard the installation by the tenant of additional moorings at a marina (at a cost of £360,000 to the tenant). Rectification was possible only because evidence of the initial negotiations clearly indicated that such a disregard should have been included.

The form of the improvements disregard has been the subject of a long-running debate. The usual formula — indeed, the one favoured by the Model Form — disregards "the effect on rent" of any improvement carried out by the tenant. This is favoured because a disregard of the improvement itself might unduly favour the tenant (eg where the tenant has installed vital services). While protecting the tenant against any increase in rents attributable to improvements, the "effect on rent" formula also insulates the landlord against any reduction in rent where improvements would otherwise have that effect (eg by reducing the lettable area of the premises). In leases of large office buildings in the City of London, arguments about the premises to be valued and what improvements they may or may not include have been avoided in recent years by the inclusion of a specification of the building — usually the original specification of the building under review — which clearly shows what the tenant is bidding for.

Where a tenant has to carry out improvements to premises in order to comply with statutory duties, those improvements will invariably be carried out pursuant to an obligation to the landlord. This is because all institutional leases contain a covenant by the tenant to comply with all legislation affecting the property. An issue may arise from the Disability Discrimination Act 1995 ("DDA") which already requires a tenant to make reasonable adjustments to ensure that disabled employees (or potential employees) are not discriminated against. From October 2004 the provisions of the DDA relating to the provision of services to the public took effect, thus requiring businesses to ensure that there is no discrimination against disabled members of the public in the provision of services. Compliance may manifest itself with, say, the installation of a wheelchair ramp where there was previously a step outside a shop. Arguably, this would be an improvement carried out pursuant to an obligation to the landlord and so would not be disregarded at rent review. However, some commentators discount this analysis. Their argument is that the DDA is not a property statute. Duties under it do not relate directly to the physical attributes of premises but to the provision of services. A business could make arrangements that did not involve alterations to the premises from which it operated — eg by arranging for the delivery of goods or for the provision of services at other premises. In this analysis the decision whether or not to make alterations to the premises would be entirely within the tenant's discretion — subject of course to any constraints in its lease about the carrying out of alterations — and not the result of an obligation to the landlord.

If the rent review clause is silent on tenant's improvements or alterations, it was held in 1979 that there can be no assumed disregard but there will probably be a significant discount in rent because such a hypothetical lease would be considered onerous with the tenant not only having to pay the cost of improvements or alterations but also an enhanced rent to reflect the changes. The discount in rent will vary according to the circumstances but might be in the range of 5–10%.

Alchemy in the UK — headline rents

It is not the purpose of this book to provide a minute analysis of the wording of the RICS/Law Society Model Form of rent review clause; neither is it an academic review of the many cases that both machinery and procedure have generated over the years. However, one particular episode, provoked by the property crash in the early 1990s, is worth reciting. In our view, this episode (the aftermath of which lives on) tells us much about institutional landlords' attitudes and the court's approach to rent review.

The development boom of the late 1980s led to a market crash in the early 1990s. By way of example, prime City of London office rents moved from £70 per sq ft in 1989 to £35 per sq ft in 1993. (Even in 2006 they had only recovered to £45 per sq ft) As the vast majority of rent reviews are upwards only the landlord could look forward to 1988 rent levels after the 1993 rent review. But rent review evidence is largely based on "comparables" — similar premises in similar locations which have recently been let in the open market, thereby generating evidence of what the open market rent is at the relevant time. Landlords are always reluctant to participate in transactions which provide evidence that the market has weakened. We saw this in Chapter 4 with the consideration of the underletting condition which prohibited underleases being granted at less than the passing rent under the headlease. The landlord wants the cake and the penny. In the early 1990s, lettings were few and far between. Those that were concluded often appeared to be at rental levels similar to or greater than what had been achieved in the booming market of a few years earlier. The landlords managed this by one of two means; the payment of reverse premiums by landlords to tenants on the grant of the lease (an unpopular practice which disappeared completely after a couple of well-publicised instances of tenants receiving several such payments then promptly vanishing) or the grant of a long rent free period.

Rent free periods on the grant of rack-rent leases have always been common. Their stated aim was usually (and as contemplated by the RICS/Law Society Model Form) to compensate the tenant for the time it would have to spend fitting out the premises to enable it to occupy them for the purpose of its trade or business. In weaker markets these periods would often be extended — say, nine months instead of three. In the early 1990s very long rent-free periods were granted — sometimes as long as five years. The effect of these long concessionary periods was to inflate artificially the rent that the landlord could claim was being paid for the obvious reason that it would not be payable until a distant future date. These rents became known as "headline" rents. Of course, discounted cash flow valuation techniques would enable these "headline" rents to be reduced to a present day value, but curiously this analysis did not seem to take place. (Indeed, another feature of the letting market of the early 1990s was the widespread use of confidentiality agreements on both lettings and sales. This might have had a bigger impact on the ability of valuers to identify and correctly analyse comparables at rent review if it had mattered. For the reasons set out above, however, there were very few contested rent reviews in the early 1990s.)

So far, so strange. The landlord has an upwards only rent review so why does he need to maintain a fiction about the rentals achieved in a depressed market? We think there were two reasons. The first is the "comparables" issue referred to above. The second is the benefit on rent review under the lease granted at a headline rent. Although the early 1990s saw the beginning of the trend towards shorter leases and break rights, most of the leases were for at least ten years and many were for 15. The landlord was thus guaranteed a period after the rent free period of at least five years of the headline rent after the first rent review. If the market managed to improve beyond the headline rent by the first rent review, so much the better for the landlord.

Needless to say, headline rents introduced a (possibly unintended) consequence to the rent review process. Landlords started to issue rent review clauses in which evidence of *all* rent concessions was to be disregarded. This disregard, the landlords would maintain, would apply not only to the long-time market practice of fitting out works allowances, but also to the three- to five-year inducements which the market collapse had provoked and which were becoming the norm in the early 1990s. If the landlord's argument was correct, the tenant would be required to pay a "headline" rent at review.

The courts heard a number of cases on the topic in 1994, the appeals from which were consolidated in the Court of Appeal in 1995 including *Broadgate Square plc* v *Lehmann Brothers Ltd* [1994] 1 EGLR 143; and *Scottish Amicable Life Assurance Society* v *Middleton* [1994] 1 EGLR 150. The cases themselves generated much academic and professional comment at the time and the reader can pursue these lines separately if they are of interest. Although one case resulted in judgment for the landlord — the Court of Appeal holding that a clause requiring the rent to be such as would be payable "after the expiry of a rent free period" left it with no option but to accept a headline rent — the others were all decided in favour of the tenant either at first instance or on appeal. Even in the case where the landlord succeeded, the tenant was able to convince the arbitrator that the interpretation of the rent review provision sought by the landlord was so onerous that it would cause the "willing tenant" to reduce its bid for a lease containing that provision (which the notional lease would, being on the terms of the lease containing the rent review provision). The mental agility required to balance these conflicting outcomes in one case should not be underestimated; in a subsequent case between the same parties it was held that it would be an unfair advantage to the tenant to accept the argument for a reduction as the leases of comparable premises also contained the onerous provision on which the tenant was basing its claim.

This series of cases marked a judicial tendency to seek a commercial result on rent review, rather than penalise the tenant in circumstances where the "headline rent" arguments could be run. The flaws in asserting that a headline rent was a true comparable have already been demonstrated. This judicial tendency has been identified as "the presumption of reality".

The presumption of reality is displaced — notional lease terms

In the intervening decade the courts have leaned heavily on the presumption in favour of reality when interpreting rent review provisions. When looking at the assumed length of the hypothetical lease, judges have strained to find that it will be equal to the remaining term of the actual lease. This was originally to ensure that the tenant was not called upon to pay the higher level of rent appropriate to an institutional lease of 20–25 years at a point when his actual lease had only five or ten years to run. Significant market changes during the past decade have produced a shift towards shorter, more flexible leases. This means that directing a valuer to assume a term of 20–25 years at each review is now likely to produce a lower rent — again prompting the court to lean in favour of reality. However, we were reminded in a 2003 case that the presumption in favour of reality is merely a presumption: *Canary Wharf Investments (Three)* v *Telegraph Group Ltd* [2003] 3 EGLR 31. It is not a mechanistic rule of construction. So, if the lease unambiguously directs the valuer to depart from reality then he must do so — even if the results do not please the landlord or tenant.

The case in question was brought to establish the correct length of the term of the notional lease where the original lease term was 25 years. The landlord contended that the 25 year term ran from 1 April 1992; the tenant that it was assumed to begin afresh on each rent review date.

The judge reviewed the authorities both as to circumstances where the presumption of reality might apply and where the court had been required to supply a term for the notional lease at review. He found that there was no authority for a mechanistic rule of construction of universal application. In this case the natural meaning of the clause was that the assumed term was 25 years from the rent review date. This view was reinforced by other provisions in the lease. He therefore held that the term of the hypothetical lease for rent review purposes was 25 years from the relevant review date. He would not accept the imposition of a presumption of reality to upset this conclusion. Indeed, he thought that to do so would elevate the presumption to a "mechanistic rule".

Throughout the 1960s and 1970s, landlords of large office buildings considered that upon rent review it would favour them if the hypothetical term was as long as possible, perhaps 25 or even 35 years, as opposed to a fairly short unexpired residue, which might give rise to a discount from the full rental value of the property. During the 1980s, with American companies' preference for short leases being a particular influence, lease lengths reduced and a well-advised landlord would have been cautious as to the term assumption on rent reviews. By 1992, when the lease in question was granted in Canary Wharf, it is doubtful whether the landlord would have chosen a 25 year term to be assumed on each rent review, although a well-advised tenant might very easily have accepted this, knowing that there would have been a possibility of achieving a discount when the rent reviews came around.

Despite this case being decided in favour of the original term of 25 years being assumed at each rent review, there are other cases which remain important pointers to the assumption of the unexpired residue of the lease. In a 1990 case where the words used were "term of years equivalent to the said term", this was held to mean backdated to the beginning of the lease, with the result being that on review it was the unexpired residue of the lease to be assumed. This may be contrasted with the case analysed above where the words used were "commencing on the review date" and it was held that the full lease term would be assumed at each rent review.

In today's market, landlords are providing flexibility to tenants by granting shorter leases, very often with breaks — an approach which, one hopes, is informed by the existence of the Commercial Lease Code and influenced by the other factors examined in Chapter 9. This flexible approach is mirrored on review. For a large office building, even on the last review, five years before expiry, it is unlikely that the landlord would choose or the tenant would accept a term assumption of more than 15 years, unless there are breaks included in the assumed term. For smaller properties, almost without exception, the presumption of reality will be accepted by landlords and tenants, meaning the assumption of the unexpired residue when reviews occur. Large retail warehouses are an exception to this principle, with 25 year terms still commonly accepted and not being regarded by tenants as onerous or inflexible, unlike the situation for office tenants.

It remains possible to draft a rent review provision requiring the valuer to depart from reality but clear words must be used. It is evident that the courts will not help a party to achieve a better result by imposing reality where the clause clearly directs the valuer to assume facts which do not accord with that reality.

Upward/downward rent reviews

Since rent reviews became common in the late 1960s and 1970s, almost without exception, they have been construed on an upward only basis. This was demanded by the institutions in particular, so that commercial property offered an investment medium which was a guarantee that the investor's income would not reduce over the period of the lease. This was particularly relevant during the time when inflation was at its peak in the late 1960s and throughout the 1970s. The only exceptions to the "upward only" rule were geared leases which might have had an upward/downward provision or where there were unusual circumstances. There has been a considerable debate over the years as to the morality of commercial rent reviews being on an upward only basis and this became particularly topical over the last twenty years, in the central London office market, which has seen a succession of peaks and troughs throughout this period. An office tenant who acquired a building in central London in, say, 1990, when the market was at a peak and took a 20 year lease with five yearly upward only rent reviews, might have paid £50 per sq ft at the beginning of the lease. By the time of the first rent review in 1995, the rental value may only have been £30 per sq ft but nevertheless because of the upward only rent review provisions, would have to pay a rent which was £20 per sq ft above the market rent for the ensuing five years until the next upward only rent review in 2000.

In 2002, the Government introduced a Code of Practice for Commercial Leases, which stipulated that property owners should offer priced alternatives to commercial tenants when granting leases of properties. The Code is considered in more detail in Chapter 10. The relevance of this to the upward/downward rent review debate, is that tenants would be offered a lease with an upward only rent review at a level of rent, but as an alternative would be offered a lease with an upward/downward rent review clause, but at a higher level of rent to compensate the landlord for the possibility that the rent might actually reduce in five years time at the first rent review. The reality is that very few if any tenants would choose the upward/downward rent review at the higher level of rent and this is a classic case of tenants taking a short term view and wishing to secure the property at the lower level of rent and be prepared to face the consequences in five years time, should the rental value be lower and therefore having to continue at the inflated level of rent for a further five years. Legislation has been threatened by the Government, but to date they have steered away from actual intervention and even following the last general election success of New Labour in May 2005, this topic does not seem to be very high on the political agenda at present.

There is no doubt that if Government legislation were introduced, even to the limited extent of having upward/downward rent reviews, but to no lower than the initial rent at the beginning of the lease, this would act as a serious disincentive for institutions to wish to invest in commercial property and the dramatic effect of the upward/downward rent review clause on a large office building is well demonstrated, by giving the example of such an office building in Victoria, which has upward/downward rent review provisions and by considering the rents which have been agreed or determined at arbitration over the last 20 years.

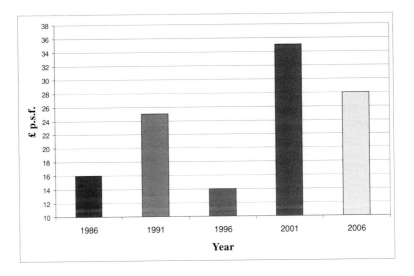

Figure 5.1

From the graph it will be seen that between 1991 and 1996, the office rent reduced from £25 per sq ft to £14 per sq ft and this resulted in an actual saving to the tenant over the five year period of approximately £9 million. A further huge saving is now being made by the tenant following the March 2006 rent review and the reduction in rent from £35 per sq. ft. to £28 per sq ft, will produce a saving to the tenant over the five year period until the next rent review in the lease in March 2011, of £5.78 million. It can therefore be seen that the tenant in these two five year periods (1996 to 2001 and 2006 to 2011) will have saved a total rent of approaching £15 m.

A further reason why we consider that legislation is unlikely, is that lease lengths have reduced substantially over the last ten to 15 years, the result being that there are far fewer rent reviews today than there were in the past and at the end of a commercial lease, whether inside or outside the Landlord and Tenant Act, the rent will be set at the market level, whether higher or lower than the passing rent.

Indexation

Unlike some countries in Continental Europe, where this is the accepted basis of rent review, in this country indexation is rare, although it is found on occasions within out of town shopping centres and also on specialist properties, where it may be difficult to ascertain the open market rent from comparables. Here a form of indexation may be used to determine the rent payable on future rent reviews, sometimes linked to the Retail Price Index.

Fixed rental increases

There has been something of a trend in recent years for landlords to choose fixed rental increases in leases, rather than a rent review on an upward only basis to the open market rental value. This has the advantage to the investor of certainty throughout the term of the lease, which also is an advantage to the tenant in occupation, but would often be frowned on, due to the fact that in theory, the rental being paid by a tenant in a market place, should bear some resemblance to the strength or otherwise of the property market. A variation of this will be a fixed increase to a certain amount or the open market rental value, whichever is the higher, perhaps with an upper limit to the level of rent which will be paid. Even so, today, the vast majority of leases of commercial property still contain upward only rent review provisions.

Dispute Resolution

The two principal forms of dispute resolution found in institutional leases of commercial property are referrals to arbitration or independent expert. There is usually the ability for the parties to agree a private appointment of either an arbitrator or independent expert, but more often than not, the parties are unable to agree on an individual themselves and as a result the President of the Royal Institution of Chartered Surveyors will appoint an arbitrator or an independent expert from their own lists. These arbitrators or independent experts will be experienced private practitioners who will have been rigorously trained and are periodically subject to re-training.

There is often much debate as to whether it is preferable to choose an arbitrator or an independent expert, and much will depend on the state of the market at a particular time. Generally, in a rising market, it is preferable from a landlord's view point to have an independent expert and in a falling market, from the landlord's perspective, it is normally preferable to have the appointment as an arbitrator. This is because of the natural tendency of the rent review market to lag behind the letting market and often at a particular review date, all the transactional evidence is not necessarily generally known, meaning that sometimes the evidence which is being used is slightly historic. For this reason in a rising market, it is more likely that the independent expert, who will be able to use his own evidence and knowledge of the market, as opposed to the arbitrator who is confined to the evidence produced to him, will fully reflect the strength of the market at a particular time. In recent years, it has become a tendency within the rent review clause for landlords to be able to choose either an arbitrator or an independent expert at the relevant time, dependant on the strength or otherwise of the market. The well advised tenant at the beginning of the lease will not accept this type of clause, as it gives the landlord a potential advantage in being able to choose whether to opt for the arbitrator or the independent expert at the particular time. The choice should be made at the beginning of the lease and a general rule is that, for large buildings where the valuation issues may be complex and difficult, it is preferable to go down the arbitration route, while for more straightforward cases, independent expert would be preferred, although these rules are not always followed and different landlords seem to have different views on whether arbitration or independent expert is the better form of dispute resolution.

Arbitration

This is governed by the Arbitration Act 1996, which replaced the Arbitration Act 1979, which in turn had replaced the Arbitration Act 1950.

It is outside the scope of this book to go through the detailed process of arbitration, but the arbitrator's role is essentially to act in a quasi judicial capacity, receiving evidence from both sides, usually in the form of written representations and counter representations, before considering the matter, weighing the evidence, and producing an award, which unless the parties agree otherwise, will always contain full reasons.

Oral hearings are very rare and only occur either on very large rental disputes, or where there are legal or technical issues, and in these instances, it is normally the case that solicitors will be involved, probably with junior or even senior counsel, representing the particular parties. There may also be other professionals involved such as quantity surveyors or mechanical engineers to give evidence on any technical issues.

It is also possible for an arbitration to proceed with one of the parties not being represented and in this type of situation, the arbitrator will judge the matter purely on the written evidence, produced by one side and effectively he will deal with the matter *ex parte*.

One of the major changes which was made as a result of the most recent Arbitration Act 1996, is under section 34(2)(g) where it says that the arbitrator can take the initiative in ascertaining the facts and the law. It may be the fact that the arbitrator is aware of a piece of evidence relating to a comparable property, which has not been produced to him in the parties written representations and counter representations and in this situation, where he considers the comparable may be of significance, it is open for the arbitrator to provide the details of the comparable to the parties, and ask for their comments and this will as a result be taken into account by the arbitrator in arriving at his award. What an arbitrator should not do is to take the evidence into account without giving the parties the opportunity to comment.

Under the Arbitration Act 1996, all costs are within the jurisdiction of the arbitrator and after he has made his interim award, dealing with the substantive issue of rent, if the parties are unable to agree on the issue of costs, the arbitrator will take written representations and counter representations purely on the issue of costs and then make a final award, deciding on the apportionment of his costs as the arbitrator and also the parties' costs. Part 36 or *Calderbank* offers will be taken into account by the arbitrator, when dealing with costs, if they have been made, and this provides the parties the opportunity to have made an offer to settle the matter, which if exceeded in the case of the landlord, will provide a very favourable decision to the landlord on the issue of costs who had made the offer to settle and then as a result of the arbitration which could have been avoided, he obtains an even higher figure.

The lease may contain provisions as to how the arbitrator is to deal with the question of costs, but these provisions will be overridden by the Arbitration Act 1996 in sections 59 to 65, which give the arbitrator the full discretion on costs, whatever the lease may have stated in this respect.

Independent expert

The authority and jurisdiction of an independent expert is purely governed by the terms of the lease and there is no legislation which sets out the powers and duties of an independent expert.

The procedure may be very similar to that of an arbitration, where the parties make written submissions and counter submissions, and the expert will then take these away with him, and prepare his own findings and ultimately his independent determination. The major difference, however, is that the independent expert may decide to determine a rent which is even outside the parameters of the parties written submissions to him. This an arbitrator cannot do and the independent expert is there to make his own determination of the rent for the property in question, perhaps assisted by the parties' written submissions and counter submissions, but free to take into account whatever he wishes, and whatever evidence he knows about, in order to arrive at his decision.

The independent expert, unless it is stipulated in the lease, is not bound to give reasons for his decision, but there has been a trend in recent years due to the wish for accountability and transparency for independent experts to provide reasons or an explanation setting out the basis of the decision, as opposed to purely issuing a determination with the figure, without any explanation whatsoever as to the methodology in arriving at the figure. The downside of this is that the expert may open himself up to a claim for negligence which would be virtually impossible to pursue if the determination was purely a figure without any form of explanation.

Many leases contain provisions as to the independent expert perhaps having to make his determination within three months of the date of his appointment by the RICS, with perhaps the parties having to make their written submissions within four weeks of the expert's appointment and counter submissions within a further two weeks. These timing provisions are to be avoided, often arbitrators or independent experts are merely applied for and appointed by the RICS, as a means to parties bringing "matters to a head" and, more often than not, the arbitrator or independent expert is placed on hold after his appointment while the parties make further efforts to settle the rent. Often this period after the arbitrator or independent expert is appointed can be quite long. In some situations it can be up to two years after an arbitrator's or independent expert's appointment before he is told that the parties have finally settled the matter by negotiations and he is as a result not required to determine the dispute.

The costs of the independent expert are again purely governed by the terms and conditions of the rent review clause, and if the independent expert has jurisdiction on his costs and perhaps the parties' costs, he will have to deal with this matter in a judicial capacity, similar to an arbitrator probably by receiving written submissions and counter submissions from the parties, after determining the rent and then make his final decision on costs.

There is a great deal of similarity between the process of arbitration and independent expert, and when the draftsman, in conjunction one hopes with his client and also the client's retained surveyor, makes the decision as to whether to choose arbitration or independent expert at the commencement of a lease, it is important to bear in mind the main differences between the two processes (Table 6.1).

Capacity of appointment: interpretation of lease or agreement

Some leases and agreements may be unclear or ambiguous as to whether the appointment is in the capacity of arbitrator or independent expert. Where a lease or agreement, with reference to the appointment of a surveyor, mentions "arbitrator" or "arbitration" or "the Arbitration Acts", even though it may also make reference to a "valuer", "independent expert", "expert" or other such term, it is generally treated as calling for the appointment of an arbitrator, unless it is clear that the parties intend otherwise. An appointed surveyor should resolve any ambiguity concerning the capacity in which his is appointed before proceeding.

Other forms of dispute resolution

Mediation

This is becoming a very popular form of dispute resolution and there are several factors that make mediation different from most other forms of dispute resolution.

No decision or determination can be imposed upon the parties by the mediator, nor will the mediator express any personal view on the dispute unless the parties so request (which often occurs when there is an impasse). It is the parties' dispute and, while the mediator manages the framework of the mediation, the parties retain responsibility for the resolution of the dispute.

During a mediation, the parties are able to freely discuss the strengths and weaknesses of their case and those of the other side with the mediator, without prejudicing their position should a settlement not be reached and the dispute proceeds to litigation or arbitration. Furthermore, a party can leave the mediation at any time as it attends voluntarily.

A mediator will encourage and help the parties to generate and consider their options, and to develop these into viable courses of action. In most cases, the parties themselves will suggest options and the mediator can often add to these, sometimes putting forward ways in which other people have dealt with similar situations, perhaps indicating the range of different approaches which the courts have taken, and adding new ideas which may be relevant to the particular case. The development of options is an intrinsic part of the problem solving approach that characterises mediation.

Table 6.1

Arbitrator	Independent Expert
(a) The arbitrator acts (as does a judge) only on evidence and arguments submitted to him, but he is able to draw the parties' attention to matters of which they may not be aware. He is also able to take the initiative in ascertaining facts and the law. His award must lie between the extremes contended for by the parties. He is, however, expected to use his expertise in assessing the relevance and quality of the evidence and arguments submitted to him.	(a) The independent expert has the duty of investigation to discover the facts, details of relevant comparable transactions and all other information relevant to his valuation (though he may receive information regarding these matters from the parties).
(b) The arbitrator cannot decide without receiving evidence from the parties, or from one of the parties when he is "proceeding in default" by the other, except where proceeding on his own initiative.	(b) The independent expert bases his decision upon his own knowledge and investigations, but he may be required by the instrument under which he is appointed to receive submissions from the parties.
(c) The procedure for arbitration is regulated by statute.	(c) There is no legislation governing procedure for the independent expert and he must therefore settle his own contract with the parties.
(d) A party to an arbitration can seek and (through the courts) compel disclosure of documents or the attendance of witnesses.	(d) The independent expert has no such powers.

(e) An arbitrator may not delegate any of his duties, powers or responsibilities, although he can seek assistance.

(f) In an arbitration the arbitrator can award that one party shall pay all or part of the arbitrator's fees and all or part of the other party's costs. He can also assess the quantification of those fees and costs.

(g) The arbitrator's fees can be determined by the court under the Act.

(h) There is some (albeit limited) right of appeal against the award of an arbitrator on a point of law. An arbitrator's award may also be challenged in the courts on the basis that the arbitrator did not have jurisdiction or on the grounds of "serious irregularity". If a serious irregularity is shown the court may (in whole or part) remit the award, set it aside or declare it to be of no effect.

(i) Providing he has not acted in bad faith the arbitrator is not liable for negligence.

(e) The independent expert has a duty to use his own knowledge and experience in arriving at his own decision. However, during the course of his investigation the independent expert may seek routine administrative or other assistance from any other person. This is always provided that he is in a position to vouch for the accuracy with which such tasks are carried out.

(f) An independent expert has no power to make any orders as to his fees, or as to the costs of a party, unless such a power is conferred upon him by the lease or by agreement between the parties.

(g) There is no procedure for formal determination of an independent expert's fees.

(h) There is no right of appeal against the determination of an expert, though in some very limited circumstances the court may set it aside.

(i) The independent expert is liable in damages for any losses sustained by a party through his negligence. This is so notwithstanding that the court will not interfere with a final and binding determination that he has made.

No transcript or record of the proceedings at the mediation is made, as the talks are all on a "without prejudice" basis. The parties may represent themselves or appoint a suitable person to represent them. It is essential, however, that if any party cannot be present at the mediation, one of its representatives must be given authority by that party to agree a settlement of the dispute.

The main advantages of mediation are:

i.   each party is directly involved in negotiating its own settlement terms
ii.   no settlement can be imposed upon the parties
iii. the proceedings are conducted in private and agreement can usually be reached more quickly than may be the case if one were to litigate or to have one's dispute determined by an arbitrator
iv.   the settlement able to be reached in mediation can be far wider than the legal remedies available to a court or arbitral tribunal
v.   if undertaken early on in a dispute, mediation often enables to the parties to maintain a continued business relationship
vi.   generally, the mediation costs are considerably less than pursuing the matter through the courts or through arbitration.

To date, we are unaware of any rent reviews having been decided by taking the mediation route, but we can see the possibility of this becoming an accepted method of dispute resolution in the future. It would be possible, for example, for the lease to provide for the parties' joint appointment of a mediator or the appointment of a mediator by an appointing body such as the RICS and, if the mediation fails, then the more usual route of arbitrator or independent expert determination could be the second stage of dispute resolution. In large disputes, this would be a very useful introduction, as it may be by having a swift mediation, that the very substantial costs of an arbitration, perhaps involving technical experts and members of the legal profession, could be avoided to the parties' advantage. Training here will be an issue as to date very few chartered surveyors have received training in mediation. While there are increasing numbers of solicitors who have become experienced mediators, dealing with a wide variety of legal disputes, it could be argued that they would not have sufficient knowledge of the property market to deal with disputes involving rent.

Court

The court will only become involved in a rent review situation where there is a dispute between the parties on the interpretation of the lease and the parties have agreed to have the matter determined, usually as a preliminary point, before the arbitration or independent expert referral continues. This may result in a lengthy delay and the incurring of substantial costs and in this situation the parties may as an alternative agree that the point will be decided by a legal assessor or a barrister appointed jointly by the parties, with written representations and possibly counter representations being made to the appointed individual, who will then give a binding decision to the parties on the point of law in question. This will undoubtedly be swifter and cheaper than going through the whole court process, where often it takes up to a year to obtain an actual hearing date.

Otherwise, the involvement of a court, either a county court or the High Court, will only be of relevance at the time of lease renewal under the 1954 Act. Statistically, only a minute proportion of lease renewals under the 1954 Act are actually decided in court. Many will come quite close to a court hearing, but the parties are usually able to reach a compromise, even at the steps of the court, and the major influencing factors for this are the substantial delays and potential costs of the hearing which will encourage the parties to try and reach a compromise. The fact that the decision will be that of a judge, with little experience of the property market, may also be a factor in encouraging the parties to try and reach a compromise, without risking the vagaries of a judges ruling.

Professional Arbitration on Court Terms ("PACT")

In the late 1990s, recognising the disadvantages of referring a disputed lease renewal under the 1954 Act to court, the Royal Institution of Chartered Surveyors and the Law Society launched the PACT Scheme, offering an extension to the existing Dispute Resolution Service and Arbitration Service by providing the services of surveyors and solicitors to determine lease renewal disputes as an alternative to litigation. The PACT Scheme offers an opportunity for disputes to be resolved without the necessity of going to court. The opportunity is therefore given for landlords and tenants to have the terms and rent payable under their new lease, decided by a surveyor or solicitor acting as either an arbitrator or independent expert. The appointment will be made by the RICS or the Law Society and the professionals who are available to be appointed are experienced specialists who have been specifically trained under the PACT Scheme.

The perceived advantages of the PACT Scheme are as follows:

i.   flexibility — the parties will be able to adapt the rules to their specific requirements and choose whether the appointed professional should be a surveyor or lawyer and whether that person should act as an arbitrator or independent expert

ii.   speed — delay is one of the main problems of court action

iii. lower cost — although fees will be payable, streamlined and/or informal procedures will be available, potentially reducing the overall cost, as compared to court

iv.   quality of decision — professionals have the expertise to make sound decisions on technical matters within their specialised practice areas

v.   adaptability — the scheme will not be cast in stone and can be developed and adapted according to need.

It is recognised that it may be appropriate for the court to handle more difficult legal issues. Beyond that, it is emphasised that the reference to PACT will be purely voluntary and prospective participants should take advice on whether the scheme is suitable for their particular dispute.

The main features of PACT are set out below.

i.   It was designed to deal mainly with unopposed renewals under the Landlord and Tenant Act 1954 where a determination is required to settle the terms of the lease, or the rent, or both. Valuation in many cases may usefully be based on an agreed form of lease, but where drafting is required this function will be covered by the terms of a consent order and will influence the choice of the professional who will make the determination.

ii.   The initial focus of the scheme was within the framework of the 1954 Act and court jurisdiction. It could also be applied to other property related disputes which would otherwise be the subject of litigation and which can appropriately be decided by a surveyor or solicitor — or both sitting together. Issues such as opposed lease renewals, service charges, breach of repairing covenants, insurance and other matters arising out of leases could also be dealt with.

iii. The professional will be appointed by the President of the RICS or the President of the Law Society, who will ask for sufficient information to enable the appointment of a suitably qualified person. The appointee will act either as arbitrator or as independent expert, as decided by the parties.

iv.   Reference will be made by way of consent order from the court and the agreed terms of the order will be binding and the extent of rights of appeal to the court will be a matter for decision by the parties. Appeals, if permitted would fall within the existing legal procedures. The consent order will preserve the existing right of tenants to reject a new tenancy within a specified period subject to payment of costs and effectively relinquish their right to renew the lease.

v.   An essential part of the scheme will be the exclusion of the jurisdiction of the court over the matters to be determined by the arbitrator or expert.

It was hoped that the PACT Scheme would encourage parties not to resort to the court, but the reality is that over the last five or six years since the scheme was introduced, very few lease renewals have been decided by this method. The main problem is that there is no requirement for the parties to go down this route at the end of the lease, and in our experience, even if one or both of the parties decided to go down the PACT route, there would often be considerable difficulty in agreeing the procedure, or indeed the identity of the professional. Further, because the parties were effectively in dispute in any event, often much time was wasted in trying to get the scheme off the ground and ultimately the parties resorted to the more usual procedure of applying to the court for a hearing.

It appears to us that more effort should have been made by both the RICS and Law Society in promoting the scheme, and extolling the benefits of the arrangements as opposed to going down the court route, to both surveyors and solicitors, so that much of the procedure was effectively in place, making it very easy for the parties to go down the PACT route and therefore making a substantial saving both in time and cost, in going down this route as opposed to the conventional court procedure. It remains to be seen as to whether PACT will become the normal basis for dealing with contested lease renewals over the coming years, but to date very little progress has been made in this connection.

Flip-flop approach

This is a procedure which, as far as we are aware, has never actually been used in this country in order to resolve rental disputes between parties. It has been used in the Far East and effectively requires both parties to put the actual figure which they believe the property is worth at the rent review date, to the appointed dispute resolver and the dispute resolver has to choose one or other of these figures. He is not able to choose a figure in between the two amounts. This effectively will concentrate the parties minds and in a rental dispute of an office property in central London for example, if the correct figure is in the region of £30 per sq ft, avoids the usual situation, where perhaps the landlord will argue for a very high figure of say £40 per sq ft, and the tenant will argue for a very low figure of say £20 per sq ft and in the Flip-Flop approach, the landlord might argue say for £32 per sq ft and the tenant might argue for say £28 per sq ft and the dispute resolver will have to decide on one of these two figures, effectively eliminating the parties' tendency to argue for an extreme position.

The reason why such an approach has not been adopted hitherto in commercial property, is probably the fact that this would be limiting and unconventional and parties prefer to have the ability to argue their position to the maximum extent possible, rather than being constrained in adopting a procedure as outlined above.

Nevertheless, it seems to us quite a compelling method of dealing with a rental dispute, by encouraging the parties to put forward a figure which is realistic.

Break Clauses

History

It was established in 1760 that the inclusion of an option to determine (to give a break clause its more formal name) did not render a lease void for uncertainty. Thus a lease for nine years determinable on the expiry of the third or sixth year is valid.

Until the late 1980s break clauses were rarely included in institutional leases which were, as we have seen, designed to generate a reliable income stream for 25 years. By 1993, however, the unthinkable had become a commonplace, with 15 year terms incorporating breaks at years five and/or ten, and even ten year terms with a break at year five. The breaks could be mutual or exercisable by the tenant only. Where the landlord had a break right its position was affected by the Landlord and Tenant Act 1954 ("the 1954 Act"). The tenant's right was not so circumscribed but its position could be adversely affected by the attachment of conditions to the exercise of the break. We will consider both of these aspects later in this chapter.

Construction

Break clauses are always construed strictly. For example, time is of the essence — any break notice must be served by the date or within the period stipulated. Similarly, if a notice is to be served on a party's registered office, a notice served at the premises will not work. A common problem with a poorly-drafted break clause is that it does not make clear whether the specified date is the first or last date on which notice can be given or the first or last date on which notice might expire. Where this problem arises, it is advisable to serve alternative notices covering every possible construction.

Until 1997 it was thought that this strictness of construction would apply to all aspects of a notice. Now the rule is that a notice will work if a reasonable recipient would not be misled by it. Litigation on this point has usually revolved around dates — being a day or a year (or two!) out is not necessarily fatal.

Who may operate the break clause?

Most leases will specify who may operate a break. If no-one is specified, it is the tenant's option only. The break clause will be enforceable by or against successors in title to the original parties whether or not they are expressly referred to in the lease. If, as is sometimes the case, the break right is personal to the original tenant, it will be lost on the assignment of the lease to a third party. This is so even if the assignee is an associate of or is in the same group as the original tenant. Care should be taken on this point where corporates are reorganising their property holdings. Further, a personal right is lost forever once the assignment has taken place. Reassignment of the lease to the original tenant will not revive it.

Effect on sub-tenants — break operated by landlord

Unlike surrender and merger, the common law rule is that the determination of the headlease determines any underleases. As institutional leases usually attract security of tenure under the 1954 Act, the common law position is modified — see below.

Effect on sub-tenants — break operated by the tenant

Until 1995 it was generally thought that underleases survived the determination of the head lease by the head tenant. In 1995 the Court of Appeal held that the old view was wrong and that underleases would be determined on determination of the headlease by the head tenant. A subsequent decision suggests that the parties cannot avoid this outcome contractually. As with landlord's determination, however, the 1954 Act may affect the position of an undertenant who would otherwise lose his lease — see below.

Conditional break clauses

The effect of a condition to be met by the tenant before it could operate a break clause was much litigated in the 1990s. Apart from the procedural requirement for service of a notice (which has been considered above) the two most common conditions were:

- performance of the lease covenants and
- payment of a liquidated sum of money.

Performance of lease covenants

This was a typical example of the institutional landlord giving with one hand and taking away with the other. We have already observed that break clauses were not popular with institutional landlords; indeed, they were the antithesis of what the landlord wanted, representing (in the landlord's analysis) an opportunity for the tenant to escape its long-term obligation to pay rent. Many early 1990s break clauses were made conditional on the performance by the tenant of its covenants in the lease. As with notice provisions, that condition had to be strictly performed. It is a matter of construction of the clause whether the condition had to be met at the date of service of the notice, at the expiry of the term or both. Whatever the correct analysis of that aspect of the clause, even a minor breach of covenant would prevent the condition from being met and the option would not be capable of exercise. For example, the tenant's repairing obligation would have to be fully observed and performed and the rent paid on the due dates. While most reasonable landlords would not seek to enforce minor wants of repair during the lease term, or to take action where rent was paid a few days late, either of those occurrences would entitle the landlord to defeat the tenant's attempt to operate the break clause. And, before 1996 when the tenant's privity of contract position was somewhat ameliorated, that defeat could result in a very heavy financial commitment from the tenant to premises for which, by definition, it had no further use.

By the mid-1990s prospective tenants would usually anticipate this problem in their negotiations and breaks became either unconditional or conditional only on substantial performance of the lease covenants, or absence of a material breach. The landlord would, of course, have all its usual end of lease remedies (eg regarding dilapidations) at the expiry of the broken lease.

Payment of a liquidated sum

Another common feature of break clauses was the requirement to pay a liquidated sum — often equal to six months' rent. This was designed to compensate the landlord for the prospective void after the departure of the current tenant. Depending on the drafting, it would be payable either on the exercise of the option or on the expiry of the lease. Failure to pay at the correct time would defeat the purported exercise of the option.

Interplay with the Landlord and Tenant Act 1954

Landlord's break clauses

An increasingly common feature of institutional leases is the inclusion of a landlord's break option at some time during the term. One must distinguish here between leases that are outside and those within the security of tenure provisions of the Landlord and Tenant Act 1954.

First, for leases that are outside the 1954 Act, the landlord will have a free hand and provided the requisite period of notice is given (often three or six months), it will be able to break the lease. The tenant will not have the ability to challenge the notice if it complies with the formalities required in the lease nor indeed receive any form of monetary compensation for having to leave the premises.

The situation is more complicated, however, for leases drawn within the 1954 Act. Here, the landlord will have not only to provide the requisite period of notice specified within the lease, but also to prove the reason for operating the break under section 30 of the 1954 Act. Section 30 has seven specified grounds for the landlord opposing the tenant's application for a new tenancy at the end of the lease. Five of these will generally have no relevance to the landlord's break clause taking effect during a lease. This is because grounds set out in section

30(1) paras (a), (b) and (c) relate to a tenant's substantial breach of its repairing or other obligations or persistent delay in paying the rent. These grounds for opposition may well apply at the end of the lease, but will not be used by the landlord as a means of operating a landlord's break option during the lease. The fourth ground for opposition to renewal (section 30(1) para (d)) is the landlord's ability to provide suitable alternative accommodation for the tenant. Again, this ground would probably only apply at the end of a commercial lease. The fifth ground (section 30(1) para (e)) enables the landlord to oppose renewal where it can show that the current tenancy arose from subletting out of a head tenancy, the landlord is a reversioner to the head tenancy and the combined rents on letting the holding and the remainder of the premises in the head tenancy would be substantially less than the rent for letting the whole property. The landlord in this type of situation may well require possession of the holding to let or otherwise dispose of the whole property but this is a very rarely used ground for opposition to renewal and certainly would be most unlikely to be used as a means of breaking a tenancy during a lease.

The final two grounds for opposition might be used by a landlord in operating a break clause during a lease. The first (section 30(1) para (g)) relates to the landlord's wish to occupy the premises itself for business or as its residence. However, there is a restriction here in that, to avail itself of this ground, the landlord must have owned the reversion in the property for at least five years. Thus, in the case of a landlord's option to break, if the break option were, say, three years into a lease, where the landlord had only purchased the property at the beginning of the lease, the five year rule would not be satisfied.

The most commonly used ground for a landlord to break a lease, found in section 30(1)(f) of the 1954 Act, relates to the demolition, reconstruction or substantial construction of the premises. The landlord must show an intention to demolish or reconstruct the premises in the holding or a substantial part, or to carry out substantial work of construction in the holding or part of it. In the case of substantial work of construction, the landlord must show that the work could not reasonably be done without obtaining possession.

There has been extensive litigation over the years on this ground of opposition to the renewal of a lease or on implementing a landlord's break clause and the detail of this litigation is outside the scope of this book. There must undoubtedly be a definite intention by the landlord to carry out the work, coupled with the financial means to be able to implement the scheme and also, if appropriate, a realistic prospect of planning consent being obtained, although there does not have to actually be a planning consent in place at the time the landlord's notice is served.

Compensation

This will only apply to leases within the 1954 Act. If the tenant is statutorily protected, the tenant will receive one times the rateable value of the accommodation on providing vacant possession to the landlord and this will increase to twice rateable value when the tenant has occupied the holding for carrying on business for 14 years immediately prior to the termination of the tenancy or, if there has been a change in the occupation, the successive occupiers for that period were also successors to the business. The rateable value will be ascertained at the date of service of the landlord's notice.

Tenant's break clauses

A tenant's notice operating a break clause is a "notice to quit" for the purposes of the 1954 Act. Because the tenant's ability to terminate the lease is not circumscribed by the 1954 Act (unlike the landlord's), no more is required.

In the mid-1990s some tenants seized on the operation of the break as a means to secure a downwards rent review while remaining in occupation of the premises. The mechanics were simple. The tenant would bring the current tenancy to an end by operating the break clause while simultaneously serving a request for a new tenancy under section 26 of the 1954 Act. After much academic debate and a trip to the House of Lords for two of the protagonists, it was decided that the service of a notice to quit prevented the tenant from asserting a right to a new tenancy. (This is why some leases granted in the mid- to late-1990s required the tenant to give 13 months' notice to operate the break. This would prevent the simultaneous service of a section 26 request for a new tenancy, which could not be served more than 12 months before the contractual termination date.)

Break clauses and insurance provisions

Another area where break clauses have appeared is insurance. In the traditional institutional model:

- the landlord takes on insurance and reinstatement obligations
- the landlord insures against a comprehensive list of risks with a reputable insurer including at least three years' loss of rent
- the tenant does not have to pay rent between the date of damage by the insured risk and the date of reinstatement
- the tenant reimburses the cost of insurance to the landlord.

It will be readily appreciated that, in this model, risks have been laid off to the insurer. Negotiations outside the model require allocation of risk between the landlord and the tenant. This allocation goes to the heart of what it means to own property and the balance to be struck between the long-term investor and the (relatively) short-term occupier. Unsurprisingly to anyone who has read this far, the institutional landlord's view of how any uninsured risk should be allocated does not necessarily reflect that balance.

Failure to reinstate

The institutional landlord wishes to see its income maintained throughout the lease term. Loss of rent insurance often ends after three years. As a matter of commercial reality, it is likely that resolution of any problems connected with reinstatement would be achieved within that timeframe. Either reinstatement will have taken place or the landlord and the tenant will have come to an extra-contractual agreement to close the issue. However, some tenants balked at landlords who sought to limit the rent cesser to the earlier date of reinstatement and expiry of the loss of rent insurance. Why should the tenant resume paying for unusable premises? Enter the break clause. Some landlords would allow the tenant to break if reinstatement was not completed by the end of the period of loss of rent insurance. Some would allow a break unless reinstatement had begun. Some required mutuality (although here, especially, the tenant would need to ensure that the landlord's obligation to reinstate promptly was clear). Another possible exit for the tenant may be through the application of the doctrine of frustration. It was decided as long ago as 1981 that this doctrine might apply to leases. However, the doctrine was held not to apply on the particular facts of the case where the possibility was mooted and there is no subsequent case where it has been held to apply either.

Terrorism

So far we have looked at the break clause as a remedy for (primarily) the tenant where reinstatement does not take place within a certain time for whatever reason. The next development moves us back a stage — what happens where there is no insurance cover available against a potentially disastrous risk?

The old joke about a banker being someone who lends you an umbrella when the sun is shining then asks for it back when it starts to rain is an analogy for the less amusing state of affairs with which the property market was confronted in the early 1990s. A series of terrorist attacks in central London (St Mary Axe in 1992, Bishopsgate in 1993 and Canary Wharf in 1996) left insurers with no option but to restrict the availability of cover for damage by fire or explosion resulting from terrorist attack. After the destruction of the World Trade Center in New York in September 2001 this withdrawal extended to flooding and impact damage. It had definitely started to rain, and the umbrella was being recalled.

The Government introduced an insurance scheme in 1993, broadly enabling the "missing" cover against terrorism to be purchased up to certain limits. The premiums were designated in "Pool Re", a mutual cover arrangement. Where the cost of the damage exceeded the limit of insurance, the insured had access to Pool Re for the balance. There were then provisions for further layers of funding, initially from the insurers who had participated in the Government-led scheme with any balance being paid by the Government itself.

Flooding

The growth of terrorist activity over the last 20 years has been matched by the growth of warnings about global warming. Climate change has led to rising sea levels and summer rainstorms. Widespread flooding in Autumn 2000 led to a review by insurers of the cover that could be made available in high-risk parts of the country. The Association of British Insurers' submission to the Greater London Authority in March 2002 pointed out that London's exposure to potential flood damage was greater than any other urban area in the United Kingdom. Indeed, over one half by value - £110bn — of the threatened assets are in the Thames Region. The general growth of flood- and storm-related claims over the last ten years culminated in a "bulge" in 2000 where they represented over a third of the total claims incurred. That bulge led to a decision by the ABI to guarantee cover for two years, expiring on 31 December 2002, for domestic and small business premises. Most institutional investments will have been outside this range anyway.

The upshot of these developments is that flooding, like terrorism before it (and, possibly, problems associated with composite panels, toxic mould and asbestos in the future) has become a risk for which insurance may be at best very expensive and at worst unavailable. How is this reflected in lease negotiations?

New leases

Both the landlord and the tenant must acknowledge the fact that some potential causes of damage to premises may not be readily insurable now, or may stop being so during the course of the lease term. If the lease is drafted in the conventional institutionally-acceptable way, the tenant will face a number of problems where damage occurs through an uninsured risk:

- rent cesser, which is usually confined to damage by insured risks, will not apply
- any break right included in the rent cesser provisions (see above) will similarly not apply
- the qualification to the tenant's repairing obligation — commonly phrased as "damage by Insured Risks [as defined] excepted" — will not apply and the landlord will be able to require the tenant to reinstate the premises through the tenant's repairing obligation.

Such a state of affairs is clearly unacceptable to a tenant. It will not have entered the lease expecting to become an "insurer of last resort" when the real insurer declines to offer cover. The tenant's solution lies in the inclusion of some combination of the following provisions:

- a requirement for the landlord to insure against a full range of risks while cover is available
- the extension of the rent cesser to damage by a full range of risks whether insurable or not
- the extension of the tenant's right to break to damage by an uninsured risk unless the landlord agrees to rebuild at the landlord's cost (where similar considerations to those set out above sub nom "Failure to reinstate" will apply)

- alternatively, an agreement that the lease is frustrated where the premises are damaged or destroyed by an uninsured risk.

Institutional landlords are never thrilled at the prospect either of having their income stream brought to an end or at having to pay themselves for reinstatement, either of which is possible where these protections for the tenant are included. Nevertheless, compromises have increasingly been found. Often the landlord will prefer a route which gives it the option to pay for rebuilding and keeps the tenant on the hook while it does so. In the new world of shorter leases, the problem is somewhat less pressing than it would have been if 25 year terms were still commonplace. The Commercial Lease Code (considered in more detail in Chapter 10) recommends that, where premises are damaged by an uninsured risk so as to prevent occupation, the tenant should be allowed to terminate unless the landlord wishes to reinstate at its own cost — and the landlord should bear the risk of destruction by an uninsured risk.

Pre-1990 leases

We should also ask ourselves what the position is where the leases were granted before these problems with obtaining cover against certain risks arose. There is no answer of universal application, of course — but the potential is there for some or all of the issues set out in the preceding paragraph to arise. If the definition of "Insured Risks" includes terrorism and/or flooding and insurance is available, albeit at great cost — which the tenant will invariably have to pay — there seems to be no reason in principle why the landlord should not recover that cost from the tenant. In this situation, however, the tenant may find some consolation on rent review because of the onerous nature of the provision. And if insurance is difficult or impossible, that fact may be the catalyst for a discussion between the landlord and the tenant about the reallocation of insurance responsibilities in the realm of damage by uninsurable/uninsured risks. The parties might usefully bear in mind the Commercial Lease Code proposals referred to above when having that discussion, particularly as the context of the discussion is most likely to be a lease with no more than ten years to run.

Valuation consequences of break clauses

The institutional landlord's object is to secure an income stream for as long as possible. The property valuation process historically "rewards" (ie places a higher value on) a longer rather than a shorter term during which that income stream will be maintained. The tenant's perspective is different. It used to be offered a 25 year lease as a matter of routine. How many businesses know what their property requirements will be in five years' time, let alone 25? How many employees will it have; where will it want or need to be located? In some businesses the problem is even more acute. Some retailers have a "strategic" plan involving a one-year forecast and a "tactical" plan which runs only six to eight weeks. Yet they have to be committed to premises for a large multiple of these timeframes. Research for the RICS (Crosby/Gibson, 2001) brought home the tenant's concerns graphically. Its prime concern was lease length, followed by ease of departure (including breaks). The growth of serviced offices, property outsourcing for some government departments and larger corporates and the short-term flexible leasing offered by companies such as workspace all demonstrate that a tenant can achieve flexibility if it needs to, and is willing to pay the price to achieve it.

But to return to the institutional landlord. It used to expect a 20 or 25 year term with no breaks. Now it would settle for a 15 year term with no breaks. Any lender will look for income cover over the loan term and the innovative financing structures which have developed over the last ten years are predicated on long leases and upwards-only rent review. On imaginary premises of 100,000 square feet, a seven per cent yield and a rent of £54 per square foot, a conventional valuation would result in a value of £77.1m. The introduction of a break option will push the yield out by one or two per cent which will reduce the capital value in this example by between £9m and £17m. However, this analysis is simplistic and potentially misleading. This is because conventional valuation always assumes the worst:

- the break clause will be operated
- there will be a one year void
- the next tenant will require a six month rent free period
- there will be no rental growth when the premises are re-let.

Yet statistics gathered about leases with break clauses simply do not bear this out. The Strutt & Parker/Investment Property Databank

Lease Events Survey 2005 shows that tenants do not exercise break clauses in the large majority of cases. For example, in 2004 only 30% by rental value were exercised. Of these, reletting was achieved within a year in a quarter of the cases sampled. The highest sector was offices where 36% of break clauses were exercised. In the retail sector, this figure dropped to 18%. Even the relatively low total figure of 30% is the highest since the survey began in 1988; in 2000 tenants operated only 16% of break rights available to them.

Further, the market environment will affect the valuation of the break clause. In a rising market, the break may increase capital value if the location is a valuable one, even above that of a long lease. This is because the property would be readily lettable at a market rent, possibly to a stronger tenant. However, as we have seen, it is most likely that the incumbent tenant will not break the lease anyway; and even less likely in a rising market where alternative premises are unlikely to be cheaper. In a falling market the break clause will put the valuation at a discount to a longer lease. The risks of a long void, reduced rent on reletting even with a rent-free period and having to accept a weaker tenant all go to reduce value. However, valuation is about assumptions and probabilities, not just the application of worst-case scenarios. The increasing use of discounted cash flow (DCF) techniques in valuations and the application of portfolio theory to the institutional landlord's holdings change the perception of leases with breaks. Further, scenario-modelling software is widely available and enables a more accurate assessment of the value, by reflecting the true likelihood of the tenant leaving rather than simply assuming that it will because it can.

Break clauses versus options to renew

As we shall see in Chapter 9, there is a potentially significant Stamp Duty Land Tax saving in some circumstances where the tenant takes a lease for five years with an option to renew for a further five as opposed to a ten year lease with a break at year five. The tenant may feel that, as there is no practical difference between the two but an SDLT advantage with the former, that is what it would like. Yet institutional landlords prefer the ten year lease with a break option. Why would this be?

First, we have already seen that few tenants break. While we are unaware of any research devoted solely to the take-up of renewal options, there is certainly some about renewals at lease expiry. It seems that more leases are not renewed than there are breaks operated. This is probably explicable at least in part by a property quality issue. The leases that are coming to an end are usually of less modern premises than those where breaks might be operated. So, if research shows that more leases remain unbroken than are renewed, that goes some way to explaining why landlords prefer a break clause to a short but renewable lease. However it is at least arguable that such a comparison is irrelevant in the context of a new lease being granted today, because the tenant's choice regarding the second five years is one of form rather than substance — the premises are the same.

Second, the interplay between rent review and option to break (or not) must be carefully documented. If the lease is drafted in such a way that the review should have been completed before the break/renewal date, time will become of the essence for the review process as the new rent is a crucial (if not *the* crucial) factor in the tenant's decision whether or not to renew. This can have an unintended benefit for the tenant too, as the right to review can be lost if it is not concluded before the date at which the break can be operated — as to which, as we have seen, time is always of the essence.

Third, a series of short leases is seen as more expensive for both parties than a longer lease with a break right. This may be an overstated problem in the context of an option to renew where the terms of the lease will be set out in the lease being renewed. It is nevertheless true that action has to be taken to create and execute the renewal lease, and there will be a SDLT implication (see Chapter 9).

Finally, there is a technical risk with an option to renew. If it is incorrectly drafted so that the renewal lease itself includes the option, this perpetually renewable lease will take effect as the grant of a term for 2,000 years which cannot be brought to an end by landlord's notice. So the institutional landlord's instinct is that if a tenant wishes to leave, a break is more difficult to operate than a renewal with the result that, at the margin, a tenant is more likely to stay if it has a break than an option to renew. And, although valuation is not about the universal application of worst-case scenarios — see the previous section — the landlord's sense is that a break clause is less likely to lead to a void or the relatively high transaction costs of a renewal.

Lease Renewals

Following the end of World War II and against a background of bomb damaged sites, a shortage of accommodation and a growing recognition of the profit which could be made on the redevelopment and re-letting of property, the Landlord and Tenant Act 1954 was enacted.

Part I one provides limited security of tenure to tenants of residential property on the expiry of long leases and Part II relates to business tenancies. Earlier legislation (repealed by the 1954 Act) had provided for compensation in a limited range of circumstances for a business tenant suffering loss of goodwill on the expiry without renewal of a lease of business premises.

The 1954 Act provides a comprehensive code to balance the rights of the landlord who wishes to resume occupation, or for other reasons, to object to the grant of a new lease with those of the tenant who would like a renewal by imposing a system which provides:

i.   limited security of tenure for the well behaved tenant
ii. a framework for settling terms fair to both parties on a renewal
iii. compensation for a well behaved tenant who is refused a renewal.

In essence, the Act remained largely unchanged, with some subtle changes made within the Law of Property Act 1969, but after 50 years, following much debate, it was decided that reform was necessary to modernise the detailed operation of the Act and make the process of the renewal of business tenancies quicker, cheaper, easier and fairer.

After much debate through Parliament, the changes were finally enacted in the Regulatory Reform (Business Tenancies) (England and

Wales) Order 2003 which took effect on 1 June 2004. It was hoped that the reforms would encourage the parties to negotiate before applications were made to court for new tenancies and that this would result in savings to businesses of approximately £6.5 m in court costs alone each year.

Once again, it goes beyond the scope of this book to go through the changes in detail, but we will summarise the main changes.

1. **Contracting out of sections 24 to 28 of the Act**

   A new procedure has been set up, in substitution for the parties jointly applying to the court for an order under the previous regime. There is now a new notice procedure and the court is no longer involved. The most important aspect of the new procedure is that it must be completed before the tenant becomes contractually bound to take the lease. It will no longer be possible to have an agreement for lease which is conditional upon complying with the new requirements after exchange. The purpose behind the change is to ensure that the tenant is aware of the consequences of entering into a lease excluded from the Act, while avoiding the need to obtain court approval. If the tenant is happy to take an excluded lease, it must make a statutory declaration to that effect. The notice must contain a "health warning" that the tenant is giving up his renewal rights and advises the tenant to take professional advice.

2. **Surrenders/agreements to surrender**

   The position prior to the reforms coming into force was that a surrender was valid provided the tenant had been in occupation for at least one month, but an agreement to surrender was void unless approved by the Court. After 1 June 2004, all surrenders are valid no matter how long the tenant has been in occupation of the premises.

3. **Section 40 notices requesting information**

   This change was intended to improve the smooth operation of the renewal and termination procedures and to ensure that information supplied by one party to the other, is up to date and accurate. The reforms provide the party seeking information with a proper statutory remedy in the event that the information provided is incorrect or out of date or where no response is received at all.

4. **Wider rights — ownership and control of the landlord's and tenant's business**

    The Act makes it clear which business entities enjoy rights or are subject to obligations. The reforms aim to remove anomalies and extend the scope of the Act to cover situations where a party organises its affairs so that its property and its business are in separate ownership. While in principle the provisions of the Act apply only to the legal entities named in the lease, the reforms extend the Act to other entities under the same effective control, for example a company owned by the landlord or the tenant. It is considered that the effect of the Act should generally be neutral when decisions are made whether to incorporate a business or vest property in a company. The group company definition has also been widened so that companies are now treated as members of the same group where they are effectively controlled by the same person. This follows the general principle of looking at overall effective control, rather than at specific legal relationships.

5. **Termination by tenant**

    There was some uncertainty as to whether a tenant who wished to vacate the premises on the contractual expiry date needed to serve three month's notice pursuant to section 27 (1) of the Act in order to prevent a continuation tenancy arising under the Act. It was thought that if the tenant failed to give such a notice, the tenancy would continue until such time as the tenant gave three months' notice expiring on a quarter day. The position was clarified by *Esselte AB* v *Pearl Assurance plc* [1997] 1 EGLR 73. In that case it was held that where a tenant is out of occupation by the contractual expiry date, the lease will terminate upon that date. The reforms clear up any uncertainty by confirming the position in *Esselte*. Landlords should be very careful to check whether the tenant is still in occupation of business premises on the contractual expiry date of the lease, if formal notices have not been served.

    By virtue of a new section 27(1A), the Act now expressly avoids a continuation tenancy where the tenant is not in occupation of the premises on the contractual expiry date. If the tenant is in occupation beyond the contractual expiry date, he still has to give three months' notice but such notice does *not* have to expire on a quarter day. The fact that the notice no longer has to expire on a quarter day is of great benefit to tenants. The pre-reform position was seen as a trap for unwary tenants who, having just missed a quarter day, found themselves having to serve nearer to six months notice to terminate the tenancy, thereby remaining liable for rent and service charge for a longer period.

6. **Section 25 notices**

    Previously, the section 25 notice served by the landlord did not have to specify the terms proposed for the new lease. Now, non-hostile section 25 notices must set out the property to be comprised in the new tenancy, the duration of the term, the rent payable, the other main terms of the new tenancy and state also that the proposals are open for negotiation and that they are not legally binding. If the section 25 notice is hostile, it must set out the grounds of opposition in more detail and a statement to the effect that the tenant can challenge them.

    Tenants are no longer required to serve a counter notice in response to the landlord's section 25 notice. Conversely, landlords are still required to serve a hostile counter notice to a tenant's section 26 notice if they are intending to oppose the renewal.

    The real significance of these changes is that the landlord who is serving a non-hostile section 25 notice, possibly up to 12 months before the actual contractual expiry date, will have to propose a rent payable for the new lease at a very early date. In a rising market, this may create some difficulty. The landlord can of course cover the situation by allowing for a further increase in rental values between the time of service of the section 25 notice and the actual expiry date of the lease, although it is doubtful whether the landlord will actually be prepared to set out a rent in the section 25 notice lower than the rental value at the time of service in a falling market.

    The fact that the tenant does not now have to serve a counter notice within two months of the landlord's section 25 notice, avoids the trap that existed previously in that if the counter notice was not served on time, the tenant would have lost his security of tenure altogether. There have been many claims made against solicitors over the years for this mistake in not serving a counter notice on behalf of the tenant on time but the abolition for the need of a counter notice may leave the landlord unsure of the tenant's plans and it will not be able to make its own plans perhaps for the re-letting of the property.

7. **Renewal and termination procedure**

Before the reforms only the tenant could apply to the court for a new tenancy. The application for a new tenancy had to be made not less than two nor more than four months after service of the section 25 notice or section 26 notice as the case may be. Either party is now able to make the application to the court for a new tenancy. In addition, the landlord is able to apply for termination without renewal where he has clear grounds for opposing a new lease. Once one party has made an application, the other party is precluded from doing so.

The existing time-limits for applications to the court have been replaced by a more flexible regime. An application to court now has to be made within the "statutory period" ie the period ending with the date specified for termination in the section 25 notice or the day before the date specified in the section 26 notice. The parties are able to extend these deadlines by written agreement provided each agreement is entered into before expiry of the statutory period.

The rationale for conferring on the landlord the right to apply to the court for a new tenancy is to put the parties on an equal footing and to enable the landlord to progress litigation in the event that negotiations are not going anywhere.

8. **Interim rent**

Interim rent is the rent payable for the period between the old lease and the new lease. Before the reforms, only the landlord could make an application for interim rent, which would be determined by the court if not agreed. The reforms have introduced a number of changes to the nature of interim rent applications, the timing and the amount payable.

Now either party is able to apply for an interim rent, once either a section 25 notice or a section 26 notice has been served. An application cannot be more than six months after the expiry of the old tenancy. The start date will be the earliest date which could have been specified for termination or renewal in the section 25 or section 26 notice.

A new method for calculating interim rent has been introduced but the old method will still apply in some circumstances. Generally the new method will apply in straightforward unopposed matters and will be the same as the rent for the new tenancy. The courts do, however, have the power to vary the interim rent and award a figure which is reasonable for both parties if either party can establish that market conditions have changed significantly since the date on which the interim rent became payable or the terms of the new tenancy have changed significantly from those in the old lease. The changes to interim rent benefit the tenant who in a falling market is now able to apply to the court to gain an early reduction in rent. The changes remove incentives for either party to delay initiating the lease renewal procedure or settling the terms of the new tenancy and the reforms also provide an incentive for the landlord not to oppose the grant of a new lease where the market rent has risen significantly since it was set under the old lease and/or where the terms of the new lease are likely to remain substantially the same.

9. **Lease Term**

The maximum term which the court can order has been increased from 14 years to 15 years, which conveniently fits in with the generally accepted five yearly rent review pattern.

10. **Compensation**

The change here relates to where the tenant may be entitled to statutory compensation having occupied the premises for more than 14 years and therefore is entitled to twice the current rateable value of the premises. Before the reforms, higher rate compensation was applied to the whole of the premises irrespective of the fact that some parts might have been occupied for less than 14 years. The reforms now mean that where parts of the premises have been occupied for different lengths of time, compensation is to be calculated for each part separately.

The changes to the 1954 Act appear to be working well, and one of the main advantages is that, previously, in any lease which was protected by the Act, the involvement of the court was inevitable as a matter of procedure. Under the new regime, the court will only become involved, if there is a genuine dispute between the parties, or either the landlord or the tenant needs to apply pressure, due to the other side's unwillingness or inability to progress the lease renewal negotiations.

Do we need the Act?

This has been a debate for many years and prior to the introduction of the changes in June 2004, there was a school of thought, that it was no longer necessary to have the 1954 Act to protect landlords and tenants and that at the termination of a lease it should be purely left to market forces to determine whether there was going to be a renewal and, if so, upon what terms.

The fact that the reforms have now been introduced seems to indicate that the majority opinion in the property industry is that the Act is still necessary, in order to provide a framework for lease renewal and, most importantly, provide the tenant with a degree of protection, against perhaps an unscrupulous landlord, who for no valid reason does not wish the tenant to be able to renew at the termination of a lease. Some might argue that it should be within a landlord's power to decide who may occupy its property, but overlooks the fact that a tenant might have made a very substantial capital investment into a property, by way of improvements. Although there is the (very rarely used) opportunity to obtain compensation under the Landlord and Tenant Act 1927, the fact that the 1954 Act still exists means that the landlord must be cautious in its approach and has to face the consequences of paying statutory compensation for disturbance, either once or twice the rateable value dependent on whether the tenant had been in occupation for 14 years or more, as a measure of discretion in making a decision on whether to offer a renewal lease to the tenant. If the landlord decides not to offer a renewal, it is left having to prove one of the specified grounds under section 30 of the 1954 Act, unless it can be established that the tenant has been a persistent late payer of rent, or has been in breach of repair or other substantial matters relating to the tenancy. These are, correctly, difficult grounds for a landlord to use to oppose renewal under the 1954 Act, but if it succeeds, there will be no statutory compensation payable to the tenant for disturbance.

Ultimately, the landlord has the ability at the outset of the lease to grant a lease excluded from the 1954 Act, if he wishes to have a totally free hand at the end of the term and to be able to do what he likes with the premises at that time. The fact is today, that most leases, particularly of small properties, are outside the security of tenure provisions of the 1954 Act and the protection of the Act is usually restricted to the larger areas of space, perhaps 5,000 sq ft or more when relating this to office property and highly rented shops, perhaps with a passing rent of £20,000 pa more. The trend for leases to be granted outside the Act has increased over recent years, also due to leases generally being shorter, with the average length of commercial leases now being only just over six years (see Fig 1.2 in Chapter 1).

Encouraging Shorter Leases

Land registration

Once upon a time ... Back to Croydon where the solicitor half of the authors did the greater part of his training to be a solicitor (or "articles" as it was known in those innocent, distant times). Croydon was also the home of one of the branches of HM Land Registry, as it was then known and had been since its introduction in the late 19th century. It is now simply the Land Registry which marks its transition from a government department to an agency. In those days the register was neither open nor computerised, as it is today. If you wanted to look at a title, you needed the authority of the registered proprietor to do so. And you were certainly not allowed to look at registered leases. In cases of urgency, it was possible to make a personal search. Sometimes, armed with the registered proprietor's authority and the fee — either 50p or £1, one forgets which — the articled clerk would be despatched to Sunley House in Bedford Park to look at a title. The desk clerk would receive the authority and fee then disappear into a room full of filing cabinets, returning with HMLR's paper record of the title — the actual register of title. One was not permitted to photocopy the information proffered, nor even to make a handwritten note in ink of its contents. (Curiously, a pencil note was allowed.) No priority attached to the inspection. How times have changed! Some years later the register became open which entitled anyone to get copies of anyone else's registered title without authority, but not at that stage of leases or mortgages referred to in the register. Despite gloomy prognostications at the time, the sky did not fall down. Later, as computer technology raced forward, it became possible to make searches and obtain copies of Land Registry entries over the internet. The Land Registry began to include the price paid for a property in the proprietorship register — a practice it had formerly observed but then abandoned during a period of rampant house-price inflation during the 1970s. The drivers for these changes were, in particular, that they could be implemented because of technological advances but predominantly, we think, that residential conveyancing was seen by consumer groups and, after pressure from them, the Government, as an unnecessarily time-consuming racket designed by and for the financial benefit of the legal profession. As we shall see, allowing Land Registry rules and procedures to be driven by the consumerist desires and requirements of residential property transactions also has consequences for the commercial property industry, not all of which are necessarily desirable.

Registration of leases before the Land Registration Act 2002 ("the 2002 Act")

Until 13 October 2003 the Land Registration Act 1925 provided for leases granted for or having an unexpired term at the date of assignment of more than 21 years to be registered. This caught traditional institutional leases but not the shorter leases which, as we have seen, have become more common with the passage of time and particularly in the last ten years. Leases for 21 years or less where the tenant was in occupation were examples of "overriding interests" — not registrable, but binding on the owners of registered reversionary interests.

The rationale for the 2002 Act

The Law Commission and the Land Registry worked together for six years to produce the 2002 Act. Their joint paper "Land Registration for the 21st Century — a Conveyancing Revolution" offered three imperatives for change:

1   the need to create the legal environment in which it was possible to conduct conveyancing in electronic form and which reflected the possibilities that electronic conveyancing could offer
2   the unsatisfactory nature of the legislation that governs land registration

3   the need to create principles that reflected the fact that registered
     land was different from unregistered land and rested on different
     principles.

The fundamental principle underlying the 2002 Act is that "the register
should be a complete and accurate reflection of the state of the title of
the land at any given time, so that it is possible to investigate title to
land on line, with the absolute minimum of additional enquiries and
inspections".

The upshot of this philosophy is that more interests became
registrable thereby reducing the number of off-register interests that
can affect registered land.

Registration of leases after the 2002 Act

Leases are the area where the 2002 Act has had the most impact. Since
13 October 2003 the following interests have been registrable:

* leases granted for more than seven years and
* assignments of leases having more than seven years unexpired.

This was a contentious proposal, bringing as it does most business
leases within the ambit of compulsory registration. Indeed, a further
reduction in the registrable length of term must be expected; the 2002
Act gives the Lord Chancellor power to reduce the term triggering
registration. Ultimately any lease which has to be in writing — ie one
for a term exceeding three years — may become registrable. For now,
leases of seven years or less remain overriding interests. If the Lord
Chancellor exercises his discretion to reduce the registrable term, that
term will reduce correspondingly.

Other changes introduced by the 2002 Act

Any document received by the Land Registry after 13 October 2003 is
potentially open to public inspection — including leases and
mortgages. This is a fundamental departure from the pre-2002 Act
position; although anyone could obtain a copy of the register, leases
and mortgages were private. The presumption of availability for
inspection can be reversed if the person lodging the document
successfully applies for it to be designated an Exempt Information

Document ("EID"). The application can be made if the EID contains "prejudicial information". This is defined as information that could cause substantial unwarranted damage or distress or be likely to prejudice commercial interests. The Registrar must designate a document an EID unless he considers the application to be without merit or detrimental to the accurate maintenance of the register. EIDs do not disappear from view entirely; an edited version is made available for inspection. At the time of writing the EID system does not have a sufficient track record for an accurate assessment of its impact to be made. The extended publicity afforded to documents by the 2002 Act does conjure some interesting scenarios.

- A tenant is taking a lease on an industrial estate, or in a shopping centre. The landlord is unwilling to concede a point in negotiation of particular significance to the tenant for whatever reason, say on the repairing obligation, or service charge liability, or alienation. The tenant can obtain copies of other leases already granted on the development to see if the concession it seeks has been given elsewhere.
- It is rent review time. A "quick and dirty" assessment of current rent levels for the area can be made simply by getting copies of recently-granted leases of premises in the vicinity.

Can the effects of an open register be avoided by consigning commercially sensitive elements of the lease contract to a side letter? The Land Registry has expressed the view that it is required by the Freedom of Information Act 2000 to reveal a complete copy of any document where a third party disputes its EID status. Can this be avoided? Certainly in the case of rent, it cannot. Although SDLT does not result in the impression of duty paid on the lease (see later in this chapter) the register itself will almost certainly show the rent reserved. Further, arrangements "off register" may not bind successors in title. It is reasonable to suppose that the Land Registry will not help parties to conceal information as the 2002 Act makes it an offence to suppress information in the course of proceedings relating to registration with the intention of concealing a person's right or claim or of substantiating a false claim. Informal indications from the Land Registry are that they will not object to side letters as a means of keeping price-sensitive information secret. Developments in this area will no doubt reflect the availability and effectiveness of the EID procedure described above.

Another consequence of compulsory registration, although not exactly a change, is that leases must contain a plan which is sufficiently detailed to enable the Land Registry to prepare the title register. Key requirements are that the plan:

- is drawn to and shows its scale
- shows orientation (ie a north point)
- is in the range 1/1250 to 1/500 for urban properties
- is not based on an imperial measurement scale
- shows sufficient detail to enable identification on an Ordnance Survey map
- shows location in relation to the highway
- shows the property clearly
- identifies different floor levels (where appropriate).

It will be readily seen that this is potentially quite onerous where, for example, a series of leases of parts of a multi-occupied building are to be granted. There have been numerous instances already of documents being rejected for not meeting these requirements. But, just as the landlord must bear the cost of preparing compliant plans, so the tenant must pay to have the lease registered. The fee is calculated on the basis of the largest quantifiable annual rent reserved during the first five years of the lease term at the time of the application.

A practical example

We close this section and the next of this chapter with a comparison of two leases:

- lease A for a term of ten years with a break option at year five at an annual rent of £100,000 and
- lease B for a term of five years at an annual rent of £100,000 with a tenant's option to renew for a further term of five years.

Land Registry fees payable
Before 13 October 2003

As neither lease is for a term exceeding 21 years neither would have been registrable so no Land Registry fees would be payable.

After 12 October 2003

Lease A — £100
Lease B — not registrable, so no fee (the term does not exceed seven years).

Stamp Duty Land Tax

Stamp Duty was introduced in 1694 and underwent sporadic change until 2003 when it was "modernised" and (so far as land transactions as opposed to those involving stocks and securities were concerned) renamed Stamp Duty Land Tax ("SDLT"). Stamp duty was a voluntary duty on certain documents relating to transfers of rights and interests in land for value. Its voluntary nature reflected an absence of direct sanction if stampable documents were not stamped; the indirect sanction was that you could not sue on a document that should have been stamped but was not. As we shall see, stamp duty was a much cheaper tax for the tenant than SDLT. Statistics revealed that stamp duty was always by far the cheapest tax to collect and administer which accounts for the Government's consistent enthusiasm for it.

SDLT, unlike stamp duty, is a tax on transactions rather than documents. It affects UK land only but applies whether the parties to the transaction are in the UK or not and whether there are documents executed within or outside the UK by which the transaction is effected. It is a self-assessed tax requiring the completion of a lengthy form of return on the grant of all leases for seven years or more, or less than seven years if SDLT is due at one percent or more (see below). The rationale for change was significantly influenced by the use of agreements for lease and leases to avoid duty. Those agreements and leases are outside the scope of this book as they were not institutional rack rent leases. However, the changes are significant for institutional leases.

SDLT and leases

SDLT is charged at one percent of the net present value of the rents receivable during the life of the lease with a nil-rate threshold of £125,000 for residential and £150,000 for commercial or mixed-use properties. The calculation of net present value can be expressed as:

$$v = \sum_{i=1}^{n} \frac{r_i}{(1 + T)^i}$$

where

r_i = the actual rent due in year 1
i = the first, second, third, etc year of the lease
n = the term of the lease
T = the temporal discount rate.

or alternatively as:

$$v = \frac{r}{(1 = T)} \left[\frac{1 - \left[\frac{1}{1 + T} \right]^n}{1 - \left[\frac{1}{1 + T} \right]} \right]$$

where r is fixed for the term of the lease and the lease term is a whole number of years.

This is a considerably more complex calculation than was necessary under the old stamp duty regime. Its outcome is partly dependent on the discount rate applicable to the calculation. That rate is nominated by the Government and has remained unchanged at 3.5% since SDLT began. This is already looking somewhat ungenerous given that the Bank of England Minimum Lending Rate has moved from 3.75% in December 2003 to 4.75% at the time of writing.

As we shall see at the end of this section, a comparison of the stamp duty and SDLT treatments of our imaginary leases A and B show great differences.

How SDLT works

Agreements for lease

An agreement for lease is to be treated as the grant of a lease where "substantial performance" of the agreement occurs. Substantial performance occurs where the tenant:

- takes possession of the property or
- pays any rent or
- pays 90% of the consideration.

In the case of rack rent leases, only the first and second alternatives are relevant.

The subsequent grant of a lease pursuant to the agreement for lease is treated as a surrender and regrant. Additional SDLT will be payable on any excess chargeable consideration. If the agreement is rescinded or cancelled, however, SDLT can be repaid.

After an assignment of an agreement for lease before substantial performance, the agreement for lease is deemed to be an agreement between the landlord and the assignee. The chargeable consideration will include any consideration paid by the assignee for the assignment.

The assignment of an agreement for lease where there has been substantial performance is a separate land transaction, the effective date of which is the date of the assignment.

Assignment treated as grant of lease

Where the grant of a lease after 30 November 2003 is exempt from SDLT (because of eg disadvantaged area relief, lease and leaseback, group relief or reconstruction relief, transfers involving public bodies or charities relief), the first assignment of that lease which is *not* exempt is treated as a grant of a lease for the unexpired term on the terms on which the assignee holds the lease with effect from the date of the assignment. Such a provision is clearly necessary to prevent the avoidance of SDLT; otherwise, a prospective landlord could grant a lease to a company within its SDLT group and claim group relief and that company could then assign to an arm's length tenant, free of SDLT. Unfortunately this anti-avoidance provision operates whenever the lease is subsequently assigned. Prospective assignees will have to enquire as to the circumstances of the grant to enable them to establish whether there is a potential liability to SDLT.

Assignments generally

As we saw in the first part of this chapter, leases with more than seven years to run become registrable at the Land Registry on assignment. The Registry insist upon proper stamping before they will accept the application for registration.

Abnormal rents

As a rule, it is only the rent during the first five years of the term that is used to calculate SDLT liability. If, however, after year five there is an "abnormal" increase in the rent, the increase is treated as the grant of a new lease in consideration of the amount by which the increased rent exceeds the rent on which SDLT has already been charged and further SDLT is charged on the excess rent. An "abnormal" increase is one which exceeds 5% plus the percentage increase in RPI pa. It applies to open market rent reviews, stepped rents and, possibly, to reversionary leases. The possible avoidance route at which this is aimed is where a lease is granted at a nominal rent for five years with a review to market at year five plus payment of a sum to compensate the landlord for the uneconomic rent it has charged for the first five years. It must be said that no landlord (let alone an institutional one) is likely to do this simply to help its tenant avoid SDLT. Even if the "compensation" for the low rent were paid at the outset as a premium, there would be SDLT consequences if it exceeded £150,000. Further, as institutional landlords are looking for a secure income and occupiers like to offset annual rent payments against annual profits and are unlikely to wish to pay an upfront premium (even if discounted to reflect the accelerated payment of the first five years' rent), this avoidance route is unlikely to be operated either.

Finally, there is further irritation for the tenant as the onus will be on it to establish whether an increased rent payable after a rent review is "abnormal" for the purposes of this definition.

Rent review

HM Revenue and Custom's website glibly states that "for the purpose of working out the net present value [of the rent payable during the term], we ignore all rent changes which take place after five years". In fact the position is slightly less clear. What statute requires is for the Revenue to ignore a rent review falling five years after a "specified date". The "specified date" is one falling within three months before the beginning of the term of the lease. In these circumstances "the first five years of the term" will be extended to the rent review date. The possibility of an abnormal rent increase, discussed in the previous paragraph, cannot be ruled out at rent review.

Breaks and options to renew

In a world where a ten year lease with a break at year five is increasingly common, the introduction of SDLT in late 2003 had an impact in the way in which that five/10 year term would be taxed. We have seen that SDLT is paid on the net present value of all the rent payable during the term. Options to determine a term are disregarded for SDLT purposes — a ten year term with a break at year five will be treated as a ten year term and the full SDLT charge will be payable at the outset. The apparently rational alternative, a five year term with an option to renew for a further five years, has the SDLT advantage that options to renew are also disregarded (although SDLT will be payable on the second five-year lease if the tenant exercises its option to renew). Yet we have already seen in Chapter 7 that landlords prefer the (in SDLT terms) more expensive ten year term with a break at year five.

So we see that, from the tenant's viewpoint, there is an SDLT saving if the option to renew is not exercised and a deferral where it is. If the option to renew is exercised, further SDLT will be payable within 30 days and the renewal lease will be linked to the original lease for the purpose of recalculation of SDLT liability.

A practical example

In this section there are four scenarios, because of the change from stamp duty to SDLT on 01 December 2003.

1 Stamp duty — leases granted before 01 December 2003

Lease A (ten year term with a break at year five at an annual rent of £100,000)

The term is more than seven but less than 35 years so the rate of duty was two percent — stamp duty payable £2,000.

Lease B (five year term at an annual rent of £100,000 with an option to renew for a further five years)

The term is less than seven years and the rent exceeds £5,000 pa so the rate of duty was one percent — stamp duty payable £1,000.

(If the option to renew were exercised, the duty payable on the second five year lease would be calculated either in accordance with stamp duty rules if the option were granted before 10 July 2003 (the date of Royal Assent to the Finance Act 2003) or SDLT if it were not.)

2 SDLT — leases granted after 30 November 2003

Lease A

Net present value of the rent is £831,660; deduct zero-rate band of £150,000 = £681,600; rate of duty is one percent — SDLT payable £6,816.

Lease B

Net present value of the rent is £451,505; deduct zero-rate band of £150,000 = £301,505; rate of duty is one percent — SDLT payable £3,015.

(If the option to renew were exercised the new lease will be linked with the original lease to determine the rate of tax payable on the new lease. In this example, if the rent for the new lease were £125,000 then, assuming the discount rate and SDLT rate remain the same in five years' time, SDLT would be payable on £564,381 with no zero-rate band — SDLT payable £5,643. So the total SDLT payable would be £8,658 but £5,643 of that would be deferred for five years.)

Accounting standards

SSAP 21

Five years ago, accounting for lease liabilities was fairly straightforward. The Statement of Standard Accounting Practice (SSAP) 21, issued in 1984, required leases of any assets used in a business to be divided into one of two types:

- finance leases — these leases substantially conferred "the rights and rewards of ownership on the lessee". They would usually be capitalised and appear on the lessee's balance sheet
- operational leases (of which the classic institutional lease is a prime example) — for these, the lessee would show the current annual rent as an expense in its profit and loss account.

The idea of SSAP 21 was to recognise the differences in commercial substance of a common form of contract — the lease. For, whereas the institutional lease on its full repairing and insuring terms can be seen as leaving the risks of ownership with the tenant, it does not confer the benefits. Classic indicators of a finance lease are one or some combination of the following factors:

- ownership passes to the lessee under the lease by the end of the term

- the lease contains an option to purchase at a price that is so much lower than the fair value of the asset that it is reasonably certain from the outset that the lessee will exercise the option
- the term of the lease comprises most or all of the economic life of the leased asset
- at the beginning of the lease, the present value of the payments contracted under the lease is all or nearly all of the fair value of the leased asset
- the leased asset is customised for the lessee's use to such a degree that it could not be used by anyone else without substantial modification
- if the lease can be determined by the lessee, the lessor's losses arising from that determination must be paid by the lessee
- the lessee bears the risk or takes the benefit of changes in the value of the leased asset at the end of the lease
- the lessee may continue with a second lease of the asset at a low, non-market rent at the end of the original lease.

FRS 12

Financial Reporting Standard (FRS) 12 was introduced for financial years ending after 22 March 1999. It cut down the number of matters for which provisions could be made in company accounts and introduced a requirement for specific new provisions to be included. The background against which FRS 12 was introduced saw companies making so-called "big bath" restructuring provisions which would bring all the pain into one year in anticipation of an allegedly intended and extensive expense or series of expenses. In subsequent years "restructuring" expenditure (eg redundancy, reorganisation) would be set against the provision, thereby maintaining earnings and profits in those subsequent years.

FRS 12 eradicated this practice by requiring a company to provide only against future expenditure to which it was legally and in detail committed at the year end. Further provisions were not to be made against avoidable costs or future operating losses.

The institutional lease is squarely within these requirements. The tenant's view before 1990 was, often, that a long lease was an asset. Rents tended to increase and there was no surfeit of property available for occupation. At worst, a lease could be assigned or, perhaps, sublet at a profit rent — at least until the next rent review. The early 1990s recession changed all that. Suddenly, if a tenant had surplus property it was happy to sublet on almost any terms, even at less than the (now) over-rented level it had cheerfully agreed with its landlord in 1988 (assuming it had not also cheerfully agreed not to underlet at less than the passing rent, as we saw in Chapter 4!). So, before FRS 12, an annual rental liability of, say, £200,000 offset by an annual sublease rent of £150,000 would leave the tenant with £50,000 to "lose" in each year's profit and loss account. After FRS 12, the tenant would have to provide for "onerous contracts" of which this lease/sublease situation is certainly an example. Now the tenant would have to provide for — if we assume in our example that there are ten years to go on the £200,000 pa headlease but only five years on the £150,000 underlease — the present value of payments totalling £1,250,000. And if, in the example, there were no subtenant, the provision would be the present value of £2,000,000. If, however, the original lease had been shorter...

IAS 17

International Accounting Standard (IAS) 17 changes how leases are to be accounted for from 1 January 2005. It is similar to SSAP 21 but develops some of its concepts as follows:

- it offers additional guidance on what constitutes a finance lease
- it requires a conventional lease of land and buildings to be separated out into a lease of land and a lease of buildings at the beginning of the lease (see below)
- it introduces some rules about disclosure beyond those required by SSAP 21.

The most significant change for landlords and tenants is the treatment of the "separated out" leases of land and buildings. These are to be analysed separately in accordance with SSAP 21/IAS 17 *indiciae* to establish whether they are finance or operating leases. Leases of land will, unless title to that land is to pass to the tenant at the end of the term (most unusual), be operating leases because the land has ongoing economic value. The lease of the building is more problematic and it seems more likely that this element could be classified as a finance lease, particularly if it is for a relatively long term which will tend to consume the economic life of the building.

Where an occupier has numerous leases — eg a multiple retailer —

every current lease will have to be "split" between land and buildings as described above. The resulting leases will then be categorised either as operating or finance leases. Once that step has been taken, the opportunities for recategorisation are limited. Subsequent changes to the lease agreed between the landlord and the tenant which would (had they been incorporated in the original lease) have led to a different categorisation will be reflected but will be treated as having been made at the beginning of the lease term. However, external changes (such as revaluation of reversionary interests or revised assessments of the economic life of buildings) or circumstances such as tenant default will not trigger a recategorisation.

Finally, at the time of writing it is not clear how the rental obligation arising under the "whole" lease will be split between land and buildings.

A capital(isation) idea for the future?

The Accounting Standards Board (ASB) paper, "Leases: Implementation of a New Approach" came hard on the heels of FRS 12, in late 1999. It set out the G4+1 international group of setters of accounting standards' position on the treatment of leases in financial reports. It differs significantly from SSAP 21/IAS 17. The main point of difference is the treatment of finance and operating leases — a distinction which G4+1 found arbitrary and unsatisfactory. It recommended:

- the abolition of the distinction between finance and operating leases
- that lessees should record, at the inception of a lease, the fair value of the rights and obligations conferred and imposed by the lease
- "fair value" would be the present value of the minimum payments due under the lease together with any other obligations
- lessors should report separately the income receivable from the lessee and the residual value of the leased asset because of their different risk profiles. The assets reported by the lessor should mirror the liabilities reported by the lessee.

G4+1 observed that the "unusual" property leasing structure in the UK would make the proposals of especial interest in the UK.

Although the original closing date for comments on the ASB's paper was 07 April 2000, the debate has rumbled on. It is generally accepted that the changes proposed by the ASB are inevitable. However, the balance between the ASB's concern that liabilities (in the form of commitments under property leases) are being hidden and the accountancy/surveying professions' concerns about the impact on the property industry and the apparent financial commitments of tenants remains unresolved. So do a number of details about the accounting treatment of leases in this brave new world. For example:

- the capitalised lease obligation will have a matching asset of the same value on the balance sheet but the rate of depreciation/amortisation of the asset/liability is not known
- the liability will, as seen, be the net present value of payment obligations under the lease but it is not known whether any adjustment will be made for rental growth.

What will this mean for tenants and how will it affect their analysis of their property commitments?

- Sale and leasebacks and recently-developed variations on this theme will no longer remove assets from the balance sheet
- Non-institutional lease terms will be demanded; if landlords hold out for longer terms, tenants will seek break options as the capitalisation will, it seems, be limited to payment obligations up to the break date
- Retail occupiers may pursue "true" turnover rents. Such a rent, it is believed, would be treated as a royalty and in the absence of a contractual liability to pay any specific amount there would be nothing to capitalise and put on the balance sheet. This is unlikely to prove attractive to landlords whose idea of a turnover rent is a hybrid combining, say, 70% of market rent and a share of the tenant's turnover which will vary according to the tenant's business. Even so, this hybrid would serve to reduce the capitalised sum on the tenant's balance sheet.
- We have seen that the leisure and hotel industries prefer longer leases over which their high initial fitting-out costs can be amortised. The accelerated write-off engendered by a shorter lease would hit their profits. These classes of occupiers will almost certainly wish to maintain longer leases.
- The newly-capitalised lease obligations would be viewed as debt. For some tenants this could prove disastrous as, at a stroke, they may find themselves breaching their banking covenants by exceeding agreed gearing ratios (although the change in

accounting standards would of itself have no bearing at all on the viability of the tenant's business in cash-flow terms). The wording of the loan documentation will be crucial here. Calculation of gearing by reference to the accounting standards current *at the date of the facility* will save the tenant/borrower. The ability of the lender to revisit gearing ratios on the basis of accounting standards from time to time applicable during the life of the facility could prove fatal to the tenant/borrower, depending on the lender's desire to maintain the relationship.

The 2002 Code of Practice for Commercial Leases

Laissez-faire or legislation?

It would be an exaggeration to describe the commercial property industry as a pariah in terms of business clout and relations with the Government, although an inference of a "no-one likes us, we don't care" attitude can sometimes be drawn from the testy summaries issued by various property representative bodies after another failure to influence legislation or policy in discussions with the Government. Nevertheless, it is clear that the desires and problems of the industry are not high on the Government's agenda. Although there has been evidence in the preceding chapters of this book of Government intervention affecting the terms of the institutional lease, it is really only the Landlord and Tenant (Covenants) Act 1995 that has been directed at the perceived mischief of allowing the parties to strike whatever bargain they choose. The changes to stamp duty, land registration and accounting standards were made in spite of common arrangements for commercial leasing in the UK. Rent review, despite the existence of an RICS/Law Society model form, is still an area where a wide range of outcomes can be agreed, often to the exclusive benefit of one party (not usually the tenant).

Although the provision of premises for business is crucial to the economic well-being of the country there is no "voice for property" in the way that, say, manufacturing industry has the CBI or retailers the BRC. There is a whole alphabet of property-based or interested organisations in existence — ABI, BPF, BCO, BCSC, IPD, IPF, RICS, etc but, while all are sizeable groups with large memberships of reputable corporates, professionals and/or individuals, none could claim to speak for the whole range of those interested in property. This issue was the subject of a lively and fascinating debate at the IPD/IPF Property Investment Conference in Brighton in 2001. Peter Mandelson explained, bluntly, why the commercial property industry has a problem in making itself heard by politicians. Peter Verwer, the CEO of the Property Council of Australia, related the Australian experience where property had turned itself into a powerful lobbying group. Most who attended the Conference would have been inspired by what they heard and would, we think, have agreed that the organisation of a "voice for property" in the UK was both an urgent task and highly desirable. Nearly five years down the line, progress on this front (if indeed there has been any) remains a closely guarded secret.

Another problem for the industry's profile is its minor role in the stock market. Of the FTSE 100 companies, only Land Securities (with a market capitalisation at £9bn in August 2006), British Land (£7bn), Liberty International (£3.8bn), Hammerson (£3.6bn) and Slough Estates (£3.1bn) figure. It is unlikely that they will ever be joined in that select group as, increasingly, quoted property companies are taken private. The hostility of the stock market to property companies derives in part from the opacity of their reporting (which has improved but is inevitably hobbled by the imperfections of the property market). The quoted sector is small, and shrinking. The story goes that, a few years ago, in one day the value of shares *traded* in Vodafone (which was then the largest company in the FTSE 100 by capitalisation) exceeded the combined capitalisation of all the quoted property companies.

Against this background the reader may feel that Government would incur little political or economic damage by doing as it pleased in the areas of policy and legislation that affect institutional leases. Yet the Government has usually invited the property industry to put its own house in order where it thinks there is a problem. The subject of this chapter is a case in point.

The Commercial Leases Code of Practice 1995

Following the property crash in 1990, the Burton Report in 1992, a consultation paper from the Government in 1993 to which the industry (in the shape of the Investment Property Forum and the Association of

British Insurers) responded, a voluntary Code of Practice was introduced in December 1995. The Code's objectives were threefold:

- to improve tenants' understanding of the terms and conditions applicable to their use and occupation of business premises
- to encourage landlords to offer more flexible lease terms than had hitherto been the case and
- to increase transparency in the property market and to do away with confidentiality agreements.

The University of Reading reported to the DETR on the evidence of the first three years of operation of the Code (1996–1998). The ODPM which had come into being during that period published the report in 2000. It asked three questions, the answers to which would demonstrate whether or not the objectives had been achieved:

- had the Code been observed regularly in the property market during the period under review?
- had the changes desired by the Government been reflected in the market?
- had the Code influenced any changes in market practice?

Market conditions

During the review period the market conditions were recovering after the early 1990s recession; both rents and capital values grew in 1996–1998 with 1997 showing the best lettings market. There was one key piece of legislation in this period — The Landlord and Tenant (Covenants) Act 1995 which, as we have seen, modified the rules relating to privity of contract from 1 January 1996. This, the review indicated, had made alienation more difficult as landlords had imposed more restrictive assignment provisions (as we saw in Chapter 4). Authorised guarantee agreements were also widespread. The effect of the 1995 Act was to encourage shorter leases and the reviewers observed that the Landlord and Tenant Act 1954 mitigated the disadvantages to the tenant of a shorter lease (the effect on the landlord was not commented on). Any influence of foreign market practices was, the report thought, for the future. Tenants were certainly moving towards an analysis of their space requirements in terms of "core" and "peripheral" and an understanding that the terms of their leases should reflect the nature of their occupation. Institutional landlords, however, were not promoting a range of lease structures but were sticking to what they knew — valuations based on comparable evidence. There was, happily, some evidence that valuation of flexible leases was on the agenda for big investors, funders and leading surveyors' practices.

The review of leasing patterns

The review of leasing patterns revealed some interesting data.

- Lease lengths had fallen from the "standard" 20–25 years that was prevalent in 1990. The reductions in term predated the Code and were a function of market weakness; indeed, it was the early 1990s that had seen the greatest change. The improving market of the mid- to late-1990s had not seen a return to the 1990 status quo.
- Lease length related to the magnitude (both size and rent) of the letting rather than local factors. Secondary properties and the oversupplied office market had embraced shorter leases but retail property leases had remained relatively long.
- Some landlords had accepted repairing liabilities in the context of short leases.
- Rent-free periods were in decline; landlords no longer felt the pressure to engineer "headline: rents.
- Upwards-only rent review ("UORR") still dominated. The market had instead moved towards shorter leases (too short to include a review in some cases) and break options (which usually coincided with review dates).
- Most leases were within the security of tenure provisions of the Landlord and Tenant Act 1954.
- Confidentiality agreements had fallen out of favour.

Had tenants come to terms with their commitments? It seemed not. Many were unrepresented during negotiations and had little grasp of the lease terms beyond the level of rent. They knew that the rent would be reviewed but did not know what would happen if they were unable to agree the revised rent with their landlord.

Conclusions of the review of the 1995 code

- The code was not a factor for those tenants who did not take professional advice; unsurprisingly, landlords would not tell them about it and there was no other way they were likely to find out about it.
- Some changes that the Government desired (eg shorter leases, retention of repairing obligations by landlords) had come about but this was a result of the market rather than the code.
- Confidentiality agreements had declined in use but UORR had survived in leases of appropriate length. The Government reserved the right to legislate on these matters.
- The review was non-committal about the future of the code. While reserving all legislative options, there might be a case for abandoning the code altogether as flexibility was entering the market. The ignorance of small business tenants, both of the code and of the wider issues pertaining to their lease commitments, remained a problem.

The code is dead; long live the code!

The Government remained concerned about the issues which had driven its predecessor to introduce the 1995 Code. There was no legislation but the Commercial Leases Working Group was reconvened under the chairmanship of Philip Freedman. It comprised the Association of British Insurers, the Association of Property Bankers, the British Retail Consortium, the British Property Federation, the Confederation of British Industry, the Forum of Private Businesses, the Law Society, the National Society of Corporate Real Estate Executives (UK Chapter), the Property Market Reform Group, the Royal Institution of Chartered Surveyors and the Small Business Bureau.

Ominously for UORR the Government pledged in its business manifesto for the 2001 General Election that "upward-only rent reviews are a source of grievance to many in the business community. We will promote greater flexibility in the commercial property market."

And so the second edition of "A Code of Practice for Commercial Leases in England and Wales" was published in March and launched in April 2002. It consists of 23 recommendations — ten relating to the negotiation of new leases and 13 relating to the conduct of the parties during the life of the landlord and tenant relationship.

Key recommendations

Recommendation 5 — Where alternative lease terms are offered, different rents should be appropriately priced for each set of terms.

This is self-evidently reasonable; the issue is whether alternative lease terms *are* offered. The institutional preference for an 80+ page full repairing and insuring leviathan does not readily transmogrify into a range of terms where, say, the landlord retains a repairing obligation, the rent includes service charge or the tenant is permitted to assign without providing an authorised guarantee agreement (see Chapter 4 and below).

Recommendation 6 — Rent reviews should generally be to open market rent. Wherever possible, landlords should offer alternatives which are priced on a risk adjusted basis, including alternatives to upwards only rent reviews; these might include up/down reviews to open market rent with a minimum of the initial rent, or another basis such as annual indexation. Those funding property should make every effort to avoid imposing restrictions on the length of lease and type of rent review that landlords, developers and/or investors may offer.

Again, why not? Well, one reason is pre-existing financing arrangements which may hobble the landlord's ability to offer anything less than UORR. A landlord may be able to charge more rent for flexibility — but his funder will certainly be looking to increase the cost of finance to reflect the increased risk that any flexibility introduces to its finance arrangements.

Recommendation 9 — Assigning and subletting: unless the particular circumstances of the letting justify greater control, the only restriction on assignment of the whole premises should be obtaining the landlord's consent which is not to be unreasonably withheld. Landlords are urged to consider requiring Authorised Guarantee Agreements only where the assignee is of lower financial standing than the assignor at the date of the assignment.

What goes around, comes around. Consent not to be unreasonably withheld was the standard (either explicit in the lease or imposed by statute) but, as we saw in Chapter 4 when we looked at the possibilities opened by the Landlord and Tenant (Covenants) Act 1995, the landlord could make assignment conditions more difficult than that. Yet informal discussions with one senior fund manager indicate that, in reality, consent not to be unreasonably withheld is the *de facto* standard today, regardless of the cunning applied to drafting the conditions for assignment to be included in an immediately post-1995 lease.

Effects of the code
Investors

The code is voluntary but the implicit sanction of legislation to outlaw UORR means that institutional investors will have to balance their traditional instincts with their future interests. To comply, landlords would have to consider some combination of the following measures:

* changes or alternatives to their standard letting documents particularly with regard to lease term, rent review and repairing obligations
* introducing monitoring of compliance with the code
* developing or buying valuation software to enable accurate valuation of "non-institutional" lease terms, both for rents and capital values.

In redefining its approach, a landlord might usefully have regard to tenants' impressions of their experience in leasing premises:

* they dislike the adversarial nature of the negotiation process
* where there is a choice, they dislike the inflexibility of UK leases and will consider locating in a jurisdiction offering more congenial leasing arrangements
* they will pay for flexibility.

The challenge for the investor is to secure a voluntary, self-regulatory environment for the negotiation of lease terms. The fear is that valuation techniques will not keep abreast of the changes to leasing structures; portfolio valuations may suffer if the valuers do not find themselves dealing with familiar data.

Occupiers

The range of occupiers' experiences should not be overlooked. The small business tenants whose plight touched the Government and provoked its original interest in the code can be contrasted with a major multiple such as Boots (a company with a similar market capitalisation to Land Securities). Boots occupies over 1400 stores in the UK and Republic of Ireland with an annual rent commitment of £265m. Its capitalised leasing commitment exceeds £2bn. Boots' experience of lease negotiations since the code was introduced may be summarised as follows:

- lease lengths have fallen
- break clauses are available in secondary locations
- there is no apparent connection between lease length and breaks on the one hand and rents on the other
- the capitalisation of lease obligations will encourage shorter leases
- there is little discussion of alternative lease structures and, if there is ever to be, both owners and occupiers will have to work at this
- there is some advantage in having the muscle of a FTSE100 company in lease negotiations but
- the majority of Boots' landlords are small property companies or individuals which reduces the likelihood of a code-compliant offer or negotiation.

Boots' resources are such that it has been able to analyse the trade-off between lease length and rental value. Its findings are that, from its perspective, it makes economic sense to pay up to 133% of the rent quoted for a 15 year lease to reduce the term to five years. Contrast this with a small business tenant. It is quite possible that it will take no professional advice at all. The aspect of the lease terms negotiation that will loom largest for it is the rent. Without professional advice, or acquiring knowledge of the market and what constitute usual lease terms from somewhere else, it will be largely ignorant of the range and cost of the other obligations the lease places on it. Certainly it will be quite unable to form a view about the relative value to it of a shorter as opposed to a longer lease term.

It seems that a key issue for the success of the code is for meaningful communication to take place between the landlord and the tenant. Commonly the landlord (and certainly the institutional landlord) is aware of all the valuation issues and — subject to concerns about its financing arrangements — will know what it is willing to offer in terms of flexibility. Equally commonly the small business tenant will be ignorant of the valuation issues and the impact of flexible lease terms on the rent it should be willing to pay. It has been suggested that landlords could help reduce this ignorance by offering information, possibly supplemented by explanations eg of how alternative lease terms have been priced, to facilitate dialogue. The cynic may wonder why the landlord should agree to do this — surely it would be giving the game away? It is thought there are two reasons why such an approach would inure to the benefit of landlords as well as tenants:

- first, a change of approach seems inevitable if legislation (which has the potential to be far more damaging to investment values than a change from the adversarial nature of lease negotiations) is to be avoided
- second, the tenor of the relationship between landlord and tenant seems to be shifting from a long-term commitment having its roots in the feudal system to one of supplier and customer (see Chapter 11).

Lenders

Low interest rates and the ready availability of finance — secured on suitable investments — led to an explosion in the number of debt-backed investors in commercial property. This explosion started in the late 1990s and continues today. These investors, at the other end of the spectrum from the institutional landlord, were simply exploiting the arbitrage between the rate of interest they were having to pay on their loans and the yield of the property they were buying with it. Lenders tended to favour at least ten years unexpired on lease terms. The traditional lender analysis focussed on a combination of unexpired lease at a guaranteed (ie UORR) rent for which a creditworthy tenant was liable plus an assessment of vacant possession value at the end of the term.

Clearly the lending criteria for a code-compliant, non-institutional lease will be different. How these might be established is a matter of valuation (see below) but it seems likely that the creditworthiness of the tenant will reduce in importance. This is because the duration of the cashflow from that tenant will probably be shorter. The new key factor will be the likelihood of renewal or reletting. This, in turn, will depend on the quality of the building. Is it attractive enough to make continuous occupation and, hence, continuous cashflow a probability?

The valuation process

Rental values

In a world of full repairing and insuring leases of a uniform (or substantially uniform) length, establishing rental value is relatively straightforward. The letting agent will have some knowledge of the "going rate" for the type of premises he is dealing with. This is usually based on comparable evidence of rental values. That evidence is not always easy to come by and depends on some liquidity in the market — in short, on deals being done. Even when institutional-type lettings are the norm, this evidence is not always available.

As lease terms — particularly with regard to term, upwards/downwards rent review and allocation of the repairing obligation — become flexible, in response to the code, the range of possible outcomes for the negotiated lease expands and the likelihood of finding a direct comparable diminishes. Methodologies will have to be developed to deduce the rental uplift on a letting for, say, five years instead of 15 (as Boots has done to establish its own policies in this respect), or for a landlord assuming some repairing liability. These methodologies, once established and agreed, will themselves assist the letting agent in quoting rental terms with confidence. Unfortunately, pending their establishment and agreement, the property industry could find itself in a non-virtuous circle of no deals = no evidence = no deals.

Capital values

The letting agent's problem in establishing and justifying a rental value for a flexible lease are compounded when we consider how that lease is to be valued as an investment. The historical practice leads to the present situation where

- investors take a traditional view of what makes a good investment (long unexpired term/"clear" lease/UORR)
- valuation on a comparable basis is incompatible with the brave new world of flexible leasing
- the imperfections of the property market — in this case, almost by definition insufficient transactional evidence — prevents the valuer from carrying out his task in a way that will be clearly explicable to the client.

Ignoring the code for a moment, these ingredients lead to an environment where there is a detrimental effect on valuation yield for flexible leases. How detrimental depends on the approach of the valuer. As we saw in Chapter 7 when considering break options, valuation is not about the rigid application of worst-case scenarios to a lease. However, we also saw that the valuer will also model the cashflow on the basis that the tenant will break and that there will be a void before reletting is achieved. In probability terms, the valuer is ascribing 100% probability to each of these scenarios. And yet we also saw that over 80% of break options are *not* exercised. The valuer should really ascribe a percentage likelihood to each of the possible outcomes. So, for example, it may be that there is a 30% likelihood of a break; a 25% likelihood of non-renewal; and (in the alternative to renewal) a 20% likelihood of a six month void. Some outcomes, obviously, exclude others. In some cases the valuation will have to be on the opposite set of facts to the one set chosen by our traditional surveyor.

It is beyond the scope of this book to describe in detail the methodologies that can be brought to bear to value the variable cashflow generated by a flexible lease. Suffice it to say that progress is being made through the following approaches:

- the asset classes with which property must compete for investors' affections have themselves demonstrated an ability to price variable cashflows (eg dividends on shares, fixed and variable interest rate swaps, derivatives)
- although not widely used in the valuation profession outside the larger and specialist surveyors' firms, spreadsheet applications can quickly simulate the numerous possible outcomes for a particular set of facts
- institutional investors with large portfolios have realised that they may use the portfolio effect to advantage (see below)
- data about tenant behaviour and the leasing market is already pooled and disseminated through the Investment Property Databank
- more deals = more evidence = more deals
- cutting-edge valuation software (such as OPRent developed by Oxford Property Consultants and Fifield Software Systems) has been developed to assist investors.

The portfolio advantage

It will be readily appreciated that the effect of probability analysis described above is multiplied once it is applied to a portfolio. Put simply, the much-less-than 100% probability that a tenant will break its lease of one property, or not renew it, is reduced further still when the analysis is of ten or 100 properties; even if the worst (from the valuer's viewpoint) happens, it is highly improbable that it will happen every one of ten or 100 times.

The irony of this is that the very institutions who devised and promoted institutional lease terms are now best placed to break out of those terms and offer flexible leases, because they can spread the risk of doing so across their portfolios. The single-property-owning investor, be it an individual or a small company, with debt finance, will not feel calm about taking risks with its cashflow, regardless of the availability of sophisticated computerised scenario modelling techniques. If its tenant breaks or becomes insolvent and there is a void, it is in trouble with its lender. And even if scenario modelling did persuade the small investor that flexible lease terms could be offered, it would still have to persuade its lender that the transaction made sense in security terms. It is much safer for the non-institutional property owner to perpetuate leases on institutional terms; let the institutional owners rely on the benefits of owning large portfolios to enable them to comply with the code! And, in the time-honoured tradition of the rich getting richer, portfolio valuation is expected to be at a premium to the sum of its parts because of the overall reduction in risk to the cashflow generated by the portfolio. This advantage will never be available to the debt-backed single property player. Further, prospective tenants (assuming the quality of property they seek is available from either source) will be able to choose flexible terms from institutions or inflexible ones from small landlords, putting more pressure on the latter.

Monitoring the code

As early as November 2002 there was a rumour in the market that the ODPM had already drafted legislation against UORR — it just remained for the code to be a demonstrable failure, and it would be introduced.

This was unnecessarily gloomy. The Government had claimed on more than one occasion that it did not wish to undermine market forces. Further, a timetable for a review of the code, to be based on research undertaken by the University of Reading, was in place. An interim report, due in December 2003 was released on 20 April 2004 and a final report due in December 2004 emerged in March 2005. This timetable, by accident or design, precluded a pre-emptive legislative strike at UORR.

The interim report

This report was based on IPD and VOA data to April 2003 and a survey of 46 agents and lawyers involved in the negotiation and documenting of leasing transactions. (IPD data is generally seen to reflect the prime end of the market while the VOA covers secondary and tertiary properties too.)

The progressive findings of the interim report were as follows:

- some flexibility in the terms offered was evident, although whether this was the result of the Code or general market pressures was unclear
- lease terms were shorter — IPD data showed a reduction in new lease lengths (not rent weighted) from 16.4 years in 1997 to 13.8 in 2002
- breaks are more prevalent — 60% of leases (not rent weighted) are for five years or less or have a break option within five years
- there is a wide range of lease terms
- schedules of condition, which define and limit the tenant's repairing obligation, are increasingly used.

However, there were still three areas where limited or no progress was evident:

- the introduction of appropriately-priced flexible lease terms
- alternatives to UORR
- relaxation of assignment/subletting conditions.

Appropriately-priced flexible lease terms

We considered this issue earlier in this chapter in the context of valuation. In a nutshell, to quote Professor Neil Crosby of the University of Reading, "if you cannot price it, you cannot offer it" and the authors of the interim report expressed this opinion there. They found that there was no established pricing methodology for flexible lease terms. Even if there were, the issue would then be whether small business tenants have the skills to analyse what they were being offered — or the wherewithal or willingness to pay someone with those skills to advise them. Logically, rent should be set once the lease terms had been agreed. In practice, the authors of the report found that it was usually the other way round.

Alternatives to UORR

Quite simply, these were neither being offered by landlords or requested by tenants. This unpalatable state of affairs let to a consultation by the Government as to the options for change — see below.

Relaxation of assignment/subletting conditions

Again, the evidence was that nothing had changed. It might, of course, be argued that from the tenant's viewpoint the restrictions are less important in the context of a shorter lease, as the tenant's potential overall commitment is less. The landlord might suggest the opposite; having made the concession of a shorter lease term, it should then be able to ensure that it was happy with the quality of the income stream and its ability to prevent it from being weakened during that shorter period.

The ODPM's consultation paper and the future of UORR

Shortly after the publication of the interim report the ODPM published a consultation paper on the options for increasing flexibility in leasing arrangements with particular reference to UORR. While recognising that UORR was not wholly without merit (it ought to result in a lower initial rent for the tenant than a flexible review; it facilitates sale and leaseback transactions which can help an occupier finance its operations; it helps sustain capital values which, in turn, feeds through to the business community in loan arrangements and innovative finance structures) the Government reiterated its objections to what is, in effect, an inflation-proofing tactic in a low inflation environment:

- UORR keeps rents high in a falling market. This puts existing tenants at a disadvantage to new ones, who can pay less for the same thing. This gives the new entrant an unfair advantage in cost terms.
- Although a rack rent lease is supposed to have no capital value, in a falling market UORR can lead to negative values which make dealing difficult or, in some cases (as we saw in Chapter 4) impossible.
- The tenant takes the risk of the market falling but the landlord takes the benefit of it rising.
- The same downturn which causes rents to fall may be part of a wider economic picture which leads the tenant to need to restructure its business. Economic decline may therefore be reinforced by UORR.
- UORR can become a source of friction during a market downturn. (Perhaps this is an exaggeration; what happens during a falling market is that the landlord simply does not seek to implement the rent review — there is no point. Indeed, where the falling market is part of a wider recession there is evidence that some landlords accept a lower rent from their tenants rather than force them into insolvency.)

The options for "reform" — the property industry has consistently bridled at the prospect of legislation in this area — proposed by the Government in the consultation paper were:

1 do nothing
2 ban UORR
3 ban UORR but with a "floor" of the initial rent
4 give the tenant the right to break if UORR produced a rent above the market level
5. limit lease length and
6 require landlords to give tenants priced options.

1. "do nothing". The industry's favourite. At the end of 2003 the British Property Federation made a submission to the Government to the effect that there was no precedent for government interference in arm's length commercial contracts and that the voluntary scheme embodied in the code should be given time to work. Indeed, the consultation paper said that the Government had not definitely decided to legislate. However,

unless the code continues to be adopted more widely as some elements of the final report indicate (see below), we think continued Government inactivity is an unlikely outcome.

2 "Ban UORR". The Government's 2001 Business Manifesto certainly suggested that the continued prevalence of UORR was a particular area of concern. Barring an unprecedented wave of enthusiasm for, in effect, self-regulation by the wholehearted embracing of the code it is hard to see how this outcome can be avoided. Whether it is by this superficially simple option (would such legislation be retrospective, for example?) or one of the others is hard to forecast.

3 "Ban UORR but with a floor of the initial rent". This option seems to us to be illogical because it has the potential effect of imposing limited misery on the tenant in a falling market. Also, as we saw in Chapter 3, institutional landlords are adept at offering rental packages which inflate the initial rent to a "headline" rent. If this proposal were adopted, a "headline" rent could effectively result in the premises being overrented from the date rent became payable for the whole of the lease term. It would, however, offer some comfort to landlords and valuers by setting a minimum rent which could reliably form a component of the calculation of value.

4 "Give the tenant the right to break if UORR produced a rent above the market level". This seems highly impractical and somewhat arbitrary in its operation. By way of example: the tenant agrees the rent to take effect from the beginning of year 1. By year 3 the premises are overrented but by the review date in year 5 rental values have recovered to year 1 levels. This prevents the tenant from breaking. If the market declines subsequently, the tenant will have to pay at an overrented level again, until the next rent review or until the term ends. (Incidentally, it is probably misleading to say that UORR has produced a rent above market level — it is the market falling rather than the operation of the rent review that will have resulted in the premises being overrented. Or is the Government intending to give state backing to the presumption of reality by outlawing any attempts to direct the valuer to assume anything different on review?) If this proposal were implemented, the tenant would have to balance the costs of relocating to and fitting out new premises with paying an above-market rent for its current ones. As we saw in Chapter 7, the tenant would have no right to renegotiate a new lease at the market rent if it broke its current one as its rights under the Landlord and Tenant Act 1954

would be lost. The Government could, of course, vary that position in this context, but that would do more for the tenant (and even less for the landlord) than this proposal suggests on its face.

5   "Limit lease length". This is a blunter version of Option 4 as it would prevent a tenant from committing itself beyond an (as yet unspecified) term under any circumstances. It is hard to see how this would square with the Government's stated aim of promoting greater choice and flexibility in commercial leases. Indeed, as we have seen, tenants in the hotel and leisure sectors require the certainty of sufficiently long leases to enable them to write down their usually considerable initial fitting-out expenditure. Further, even for the majority of tenants who might prefer a shorter lease, there will be costs associated with regular renewal and an element of risk that the landlord might be able to oppose that renewal on one of the grounds allowed by the Landlord and Tenant Act 1954.

6.   "Require landlords to give tenants priced options". We already know that the mechanisms for pricing these options are neither widely available nor generally agreed upon. How could such a regime be enforced in practice? Even the consultation paper acknowledges the practical and legal difficulties. It would govern pre-contract negotiations. It would be difficult to demonstrate that a landlord had (or, come to that, had not) acted in good faith in presenting its "menu" of alternative terms to the tenant. And the group who are supposed to be the main beneficiaries of any change, the small business tenants, are the ones who most frequently do without professional advice on their lease negotiations.

The final report

The final report reviewed the same data as the interim report up to April 2004 and extended the survey to landlords, tenants (large and small), professional advisers and lenders. It was hoped that this would provide the widest evidence of how the Code had worked in practice. Both the negotiation process and its outcome were examined. Interestingly, one or two areas had shown an improvement in flexibility since the interim report. The final report found as follows.

Terms of the lease

- Lease lengths had fallen in all areas of the market to an average of between seven and eight years. However, this fall was further in the industrial and office markets than in the retail sector.
- Many lease lengths were being agreed but newer and more valuable premises commanded significantly longer lease lengths.
- Break clauses have become increasingly popular. The number of tenants exercising breaks in any year remained at 15-20% of those with the opportunity to do so.
- Schedules of condition continued to grow in use on older properties.
- Alienation provisions had not loosened. AGAs remained the norm on assignment and subletting at the greater of passing and market rent could still effectively prohibit subletting in certain market conditions (see Chapter 4).

Tenant choice

- It was still most unlikely that the tenant would be offered priced alternative lease terms. Choice was available though it would have to be at the tenant's initiative rather than the landlord's.
- Standard form leases are rarely proposed by the landlord. Some areas — particularly rent and repairing obligations — are specified at heads of terms stage, while others can be negotiated. That negotiation is more meaningful when the tenant is represented.
- Tenants negotiate, in descending order of frequency, the lease term, breaks, repairing obligations, the rent review interval and rent. How the rent is to be reviewed was rarely negotiated.

Rent review

The headline here is that, in a rising market, tenants are unwilling to pay a front-loaded rent for upwards/downwards review when the prospect of a downwards review is remote. Further, the occupier market and its advisers do not think that landlords would be willing to agree upwards/downwards rent reviews in any event.

A number of interesting statistics were revealed by the report (which relied on IPD statistics rather than VOA ones and, hence, is more telling about what the institutions are doing):

- fewer IPD leases each year contain a review — less than 50% by number but still around 75% by value at the time of the report
- if leases containing breaks are included, only 40% by number and just over 70% by value have reviews
- retail property is more likely to have reviews than offices or industrial
- the smaller the tenant, the less likely it is to have a review in its lease and the more likely it is to have a break clause
- only about 30% of offices and 25% of industrial premises have a review or do not offer the tenant the opportunity of breaking before review
- where there is a review, it is almost always upwards only to market rent
- institutional leases still adhere to a five year review pattern. Shorter review periods are the preserve of secondary property and are more likely to be found in leases of offices and industrial property than retail.

Small businesses

The corollary of the information under the previous heading is that small businesses do not occupy prime space. As a result, they can obtain shorter leases with fewer rent reviews and mitigate their repairing obligations. Against this, they have less experience of the property market generally yet do not usually seek professional advice on what they are being offered. It would be a strange tenant indeed that studied the detail of its lease but small tenants are even less well informed on this topic than the average. Nevertheless the report suggests that awareness is increasing. Agents thought small businesses were still not obtaining the best deal they could. (One hesitates to respond that they would say that, wouldn't they?) Indeed, as awareness of the code increases (see below), use of a property professional to pursue the issues on which it offers guidance may be money well spent.

The Code of Practice

Only 22% of the tenant respondents to the 2004 research were aware of the code, although nearly all of their professional advisers were aware of it. About half of them had a decent grasp of its detail and intention.

Perhaps curiously, given that the code is designed to influence negotiations rather than detailed drafting, most tenants who knew about the code had been told about it by their solicitors. A case of shutting the stable door after the horse has bolted?

The Government's response

In a statement to the House of Commons on 15 March 2005, the Parliamentary Under-Secretary of State, Yvette Cooper announced her department's conclusions on the final report and the UORR consultation in 2004. She welcomed the evidence of increased flexibility in the commercial property market, especially the trend towards shorter leases and the increased incidence of break clauses. Nevertheless, the Government remained concerned about the inflexibility of the commercial property market. She lighted on three issues:

- inflexible alienation provisions
- the prevalence of UORR in longer leases
- the lack of awareness of property issues in the small business community.

Ms Cooper proposed the following solutions:

- a review of the law relating to assignment and subletting, including legislative options
- continued monitoring of UORR but without legislation "at present"
- "action" by both Government and the property industry to raise awareness about leasing issues in the small business community
- a further joint review of the code including a new campaign to publicise it and the introduction of an effective means of dealing with complaints
- continued monitoring over the next three years.

No learning is ever wasted, they say. There is a certain irony that the development and monitoring of the Code of Practice has revealed that tenants are more concerned about the alienability of their leases than about UORR. Yet alienability is not even mentioned in the second edition of the code.

The industry acts

Just over a month after the Government's response, the British Property Federation acted through a number of its largest members. Indeed, the action had been under discussion for some months before its announcement. 20 major property companies and landlords committed themselves by a declaration dated 20 April 2005 to permit subletting in all new leases at market rent. The BPF would encourage all its members to participate in the initiative. At the same time, the signatories undertook not to enforce restrictions in current leases that prohibit underletting except at the greater of the market rent and the rent passing under the headlease. This undertaking was subject to exceptions eg objections by the landlord's lender or a superior landlord. The effectiveness of the declaration and its effect on the general market have yet to be demonstrated.

From Feudalism to Consumerism?

As the traditional institutional lease loses ground in the occupation stakes, what are the alternatives? In this chapter we consider some of the new occupational arrangements which have emerged in the last decade and a half, how property is occupied elsewhere in Europe and how the introduction of a REIT-style ownership structures might have an impact.

The RICS report "Property in Business — a waste of space?" produced in 2002 was a damning indictment of businesses' approach to their use of property and a timely reminder of the spread of ownership of commercial property in the UK. The report suggested that £18bn a year could be saved by a more efficient use of property. It is beyond the scope of this book to debate the correctness or otherwise of this figure but some of the proposals in the report about how to improve it bear examination and comparison with what has happened in practice. A key factor in future arrangements is the distribution of property ownership. The report showed that, in 2001, just over a third of commercial property in the private sector was owned by what might broadly be termed as institutions — insurance companies, pension funds, quoted property companies, overseas investors and landed estates. These are the organisations that have historically promoted the institutional lease. But nearly two-thirds is owner-occupied. The report observed that owner-occupiers use their property less efficiently than tenants (perhaps because the premises cost is not brought home to them by quarterly payments to a landlord). It is claimed that, if the owner-occupiers used their premises as efficiently as tenants, there would be a saving of "up to" £9.5bn a year. This saving is presumably predicated on the owner-occupiers selling or letting their surplus space. As we shall see, some well-established large occupiers with freehold and long leasehold premises have taken those assets off their balance sheets by a variety of methods. At the other extreme, some (usually small or start-up) businesses have avoided a commitment to property by going down the serviced premises route. The Government has completed two major outsourcings of its property holdings. The structures available to hold property portfolios are potentially changing. And in continental Europe property development — sometimes on a spectacular scale — continues against a background of shorter and in some cases state-controlled lease terms. Does the institutional lease have a future in this environment?

Serviced premises/flexible leasing

The development of these products has mirrored the growth in popularity of outsourcing. The recession of the early 1990s, apart from its effect on the property industry, raised a number of questions about how large corporations organise themselves. The concept of "core business" came to the fore. Outsourcing offered both strategic and tactical benefits. It enabled businesses to focus on their core activity, acquire high quality support from experts, offload risk (anyone who has tried to develop IT systems internally will confirm the desirability of this approach) and free resource — including space — for the core business. It also freed up capital and labour from non-core areas. Clearly, property ownership or management is not a core activity for most businesses. What options does the occupier have to meet its occupation requirements which do not involve a ten- or 15-year commitment where the only cost certainty is that the rent will go up?

The alternative providers fall into a range between completely serviced offices (where the payment covers everything and additional services are available as and when the occupier wishes to avail itself of them) and flexible leasing (where the lease terms are constructed around the tenant's requirements rather than the landlord's standard form). Neither is especially popular with valuers. Traditional valuation techniques are still applied to serviced offices and this has restricted their growth, especially in the effect of this analysis on loan valuations. Shorter leases, as we saw in Chapter 10, are probably undervalued in investment terms.

Serviced offices

The US developed the serviced offices sector originally. An early UK player was Regus, which started in 1989 in Brussels. It became the world's largest provider of serviced offices and its shares are listed on the London Stock Exchange. In common with most serviced office providers, Regus's business centres contain fully furnished offices equipped with IT and telecoms facilities; boardroom/meeting facilities are also available. This option enables the occupier to determine the extent and duration of its commitment and to vary the services taken at very short notice. Obviously the occupier pays for this flexibility but, for short-term experiments or dedicated teams whose projects have a finite lifespan, the solution seems ideal.

Flexible leasing

Flexible leasing has found operators in the office, industrial and warehouse sectors. By way of example:

- In the office sector, Land Securities offers Landflex which has been developed from its experience with LandSecuritiesTrillium (see below). Landflex enables the tenant to design its own lease or leases to facilitate expansion or contraction. Landflex also offers services — facilities management, maintenance and repair, cleaning, security, reception and utilities — and, hence, price certainty for the occupier.
- In the warehouse and industrial market, Workspace offers industrial premises on short-term lets to new and small businesses with some additional services such as broadband and preferential insurance rates. It claims that the key to its success is the active management of secondary properties and it seeks new sites through urban regeneration schemes with local authorities. At the larger end of the market, Arlington Business Parks offers a "Total Workspace Solution" where the tenant's payment includes rent, services, FM and other occupational elements but excludes only business rates. This formula has been exported successfully to Spain. Industrial and warehouse specialists Brixton has developed a product called B-Serv to enable it to offer flexible lettings of larger, newer premises in contrast to the older or secondary properties where shorter leases are traditionally available. B-Serv uses OPRent (referred to in Chapter 10) to determine the correct pricing for this flexibility. This so-called "option pricing" is of the

essence if the Commercial Lease Code is to be adhered to, yet its availability in practice is very limited. B-Serv also offers building maintenance and facilities management support to its occupier.

PFI, Trillium and Corporate Real Estate "monetisation"

Once upon a time, perhaps until the 1950s, businesses generally owned the premises from which they operated, certainly more than the two-thirds revealed by the RICS report referred to above. Around that time Charles Clore, founder of the Sears retail empire, was credited with inventing two commercial processes which are seen as commonplace today — the hostile takeover bid and the sale and leaseback. The birth and explosive growth of Sears rested on these building blocks (and Clore's early realisation that one commodity which people would always need to buy was shoes). Classically, Clore would identify a poorly-trading shoe manufacturer and retailer which owned freehold shops. The target would be taken over and the shops sold to insurance companies (often Legal & General or the Prudential) with leasebacks being taken for, in those days, 42 years. This released cash for Clore's business — perhaps to fund further hostile bids. The leases were taken on what were then institutional terms. It was a feature of sale and leaseback for the next 40 years that the occupier would agree lease terms at the restrictive end of institutional requirements to maximise the payment made by the purchaser. This makes sense when prime property generates an internal rate of return of about 7% yet most retailers could produce returns of about 17% with the capital released. Of course, an important component in the value of the investment created by the sale and leaseback is the strength of covenant of the former owner and new tenant. And the organisation with the strongest covenant in the UK is the Government.

The Private Sector Resource Initiative for the Management of the Estate ("PRIME") contract

In 1998 the Government opened the batting for the well-advised occupier by putting a contract out to tender for the management of the Department of Social Security's (as the Department for Work and Pensions was then known) 650 properties comprising 1.6 m² of space.

This came in all shapes and sizes, from large offices in city centre locations to small Jobcentres in tertiary ones. The tenure was varied, too, with a mixture of freeholds, valuable leases and rack-rented property. The winner was Trillium which was subsequently taken over by Land Securities and renamed LandSecuritiesTrillium (LST). LST's mission was to manage, fully, the 1m m^2 that the Department designated as core operational premises and to dispose of about 150,000 m^2 of space. This left about 350,000m^2 that the Department was undecided about. The deal released £250m from the sale of property, reduced the Department's occupational costs by 22% and made future costs predictable. It also saved the Department money, according to the National Audit Office — about £560m over the 20 year term of the agreement. The numerous service contracts through which facilities management was provided to the estate were reduced at a stroke to one, with LST, with a corresponding saving in staff costs.

The key difference from the traditional sale and leaseback was the basis on which the Department would occupy once the agreement was in place. It was given flexibility to move without finding a purchaser or assignee of its interest. That is now LST's job. It also has price certainty on the premises it occupies for the duration of the agreement. LST negotiates rent reviews and, through the magic of "virtual assignments" of the rack rent leases by the Department to LST, deals with the landlords (who, presumably, do not take the point that they are not legally obliged to accept rent payments from third parties). The Department shares in the savings made or income generated by disposals or development. Most importantly, the Department wanted to get out of the property business and concentrate on running the social security system.

While the details of subsequent transactions are different, the PRIME model and the processes which the outsourcing partners have applied to price the contracts offer some clues to a developing area of the property industry.

Further LST transactions

In December 2001 LST undertook two more large scale property outsourcing contracts:

- BT, where Telereal (a joint venture between LST and William Pears) entered into a 30 year "strategic partnership" with BT, covering

5.5m m² of property in 6,700 locations. Here the vast majority were freehold or long leasehold. 350 staff joined Telereal from BT. BT had run into indebtedness problems in the late 1990s (£30bn by 2000) and, while it would be an exaggeration to say that this transaction saved it, the £2.4bn raised certainly helped. Again, as with the PRIME contract, the combination of no more property risk, removal of properties from the balance sheet, quantified payment obligations, occupational flexibility and outsourced property management were seen as desirable.

- BBC. Another 30 year agreement saw the BBC transfer its White City site to LST for £37m. LST would then develop 55,000 m² of new accommodation. Other buildings would be transferred in the future as their ability to meet modern broadcasting criteria dictated. Facilities management has been outsourced as with the other contracts. The BBC joins the ranks of those favouring "one supplier, one payment".

The Strategic Transfer of Estate to the Private Sector (STEPS) contract

This followed the PRIME contract in March 2001 and related to the premises occupied by the Inland Revenue and Customs and Excise — 600-odd buildings totalling 1.4m m². As with PRIME, the properties were both freehold and leasehold and the risk of ownership was transferred to the successful tenderer, Mapeley. Mapeley's website suggests that the STEPS contract would save the Government about £1bn over its 20 year life through reduced operational costs. However, the National Audit Office reported in May 2004 that the saving would be of the order of £344m. The NAO reached this figure by comparing the likely charge with the Public Sector Comparator, a formula for costing services supplied by the public sector. The transaction released £220m to the Government which pays £210m pa for all their premises requirements (including facilities management at a number of premises which were not included in the original transaction).

Sainsbury's

The food retailer was one of the first UK corporates to examine the potential of fund raising from its large and, almost by definition, prime property portfolio. It could clearly raise money from a conventional sale and leaseback and, indeed, the finance was earmarked for expansion of its core business. However, Sainsbury's was not looking to tie itself into a series of 25 year institutional leases. The financiers of the scheme were Morgan Stanley Dean Witter who underwrote £340m worth of debt secured on the portfolio. Sainsbury's received 100% of the value of the properties and agreed long-term occupation arrangements at fixed 1% rental uplifts, while retaining an interest in any realised increase in the value of the properties. The lease terms were "soft", allowing Sainsbury's to upgrade, extend or redevelop the premises during their term without increasing Sainsbury's rental liability. Finally, the concept of a "substitutes' bench" was introduced, enabling Sainsbury's to swop properties in and out of the original group of 16 as long as the overall investment value is not diminished. So well did this scheme work that another ten supermarkets were processed in the same way in July 2000, raising a further £226m. The interesting factor here is that the investment value was created not by institutional leases but by a rigorous analysis of the cashflows generated by Sainsbury's new occupational arrangements, underpinned by the value of the portfolio and Sainsbury's covenant. From Sainsbury's point of view, while the transaction would not insulate it from property market risk (it continued to own many other stores, both freehold and leasehold), it released capital and fixed its rental commitment on the stores selected for these transactions.

"Columbus" — the Abbey National transaction

This transaction, which moved from acceptance of Mapeley's offer to completion in seven weeks, took place in September/October 2000. It was not a complete outsourcing in the manner of the STEPS contract as Abbey retained maintenance and facilities management functions, but in all other respects it reflected that transaction. 1300 properties comprising about 650,000 m² were transferred to Mapeley for £457.5m. The 20 year contract bore an initial cost of £90m per annum with an annual increase of 3%. It gives Abbey the ability to move out of half of the premises during the term of the contract and the possibility of even more flexibility for a further payment. Abbey's desired outcome here was the creation of an occupational portfolio which could be changed easily to reflect its operational requirements from time to time.

Other corporate spin-offs

The transactions listed above are probably the landmark deals in this arena. Other recent sale-and-leaseback deals have involved Woolworths (whose business was spun off by Kingfisher in 2001 with the properties going to a partnership between London & Regional Properties and Goldman Sachs' Whitehall fund) and Marks & Spencer (who sold a portfolio of 78 freehold and long leasehold stores to Topland Estates for £348m in 2001 with Marks & Spencer being able to vacate up to one-third of them during the initial 26 year term; rents will go up by 1.5 to 1.95% pa, depending on the level of the vacations). Topland went on to complete a similar transaction with Tesco in early 2004. All these transactions have featured some or all of the following:

- high quality properties
- investment-grade covenants
- funders with real financial clout
- investment bank advice or participation.

After an initial flurry of very large deals, the market now seems to have calmed. Whether this is because of external factors, or simply because the obvious deals have been done, remains to be seen. One FTSE 100 corporate real estate director has likened the process to "selling the family silver". Some of the participants had, in relative terms, a sideboard full of the stuff to take to the market. The challenge now will be to scale down the outsourcing offer — and associated costs — so that it can be applied to organisations with (to continue the analogy) just a few place settings. Certainly, LST and Mapeley seem keen to imply that these principles could be carried through to smaller occupiers. And, with every deal that is done, more evidence is available of the correct price for a contract for occupation on terms which bear little or no resemblance to an institutional lease.

The European experience

We have seen that much UK development has been organised and financed on the back of the institutional lease and its "bankability" as a source of income. We have also seen that the terms of a UK institutional lease have developed in a legal environment where, fundamentally, the parties can make whatever bargain they wish. This is a feature of the common law system that we have in England and

Wales (Scotland is different). Continental Europe does not share our common law tradition. Many of the EC countries have a legal code which derives from Roman Law, whose main features are listed below. However, following the liberalisation of the economies of Eastern Europe since 1989, many of the former Soviet bloc countries have been able to adopt leasing structures which are recognisable to the UK investor. After contrasting the common and civil law systems, we set out the main features of a commercial lease that one might expect to encounter in France, the Netherlands, Hungary and Poland.

Common law features

The main principles of a common law system applied to the leasing of land lead to the following:

- recognition of a leasehold estate as a property rather than a personal right; this means it can be assigned
- no restrictions on the terms which can be agreed (such as rent, rent review, alienation, length of lease); although, as we have seen in earlier chapters, some aspects of this freedom have been the subject of legislation — for example, alienation conditions and privity of contract.

As a result of this, a lease in which the financial obligations of the owner can all be laid off to the tenant, leaving the landlord with a "clear" rent, can be achieved.

Civil law features

Roman law treated the ability of one person to use another's land for a limited period as a personal right. Thus:

- the landlord is obliged to keep the property in the state it was at the beginning of the lease as the tenant has the right to use it in that way
- the tenant is consequently not obliged to contribute towards the cost of repairs
- the tenant may withhold rent if the property is defective
- there is no rent review — the lease will determine all rent at the outset

- alterations are not allowed, although this may be catered for in the contract; permitted alterations will have to be reinstated when the lease ends, though
- there will be no security of tenure beyond the lease expiry date
- no dealings with the lease — assignment, underletting or charge — will be allowed.

Typical lease provisions in France, the Netherlands, Hungary and Poland

Term

France — the Commercial Code dictates that a commercial lease must be for at least nine years, determinable by the tenant every three years, although this right can be negotiated out. Renewal is guaranteed if the tenant wishes it; the landlord can only prevent it by paying compensation. The landlord can also determine the lease on substantial redevelopment grounds every three years but, again, must compensate the tenant if it does so. This is based on a number of non-property factors, including removal expenses and compensation for loss of goodwill. A shorter lease of under two years can be granted but, if the tenant stays beyond the term date with the landlord's consent, it will become entitled to a nine year lease.

The Netherlands — Regulations divide leases of commercial premises into two types; business premises to which, broadly, the public has access (eg retail) which are regulated by Article 290 of the Civil Code and the rest (broadly, offices, industrial and warehouses) which are regulated by article 230a. Article 290 leases are usually for five years with a tenant right to renew for another five. The tenant can determine the lease at year five or year ten by one year's notice, but not otherwise. If the tenant holds over at expiry the lease continues until further agreement. A lease granted for less than two years ends on expiry without the need for notice. Article 230a leases are not regulated but are usually granted for five or ten years with an option to renew for a similar period.

Hungary — there are no limits on duration or guarantees of renewal. Leases are often granted for up to five years with a tenant's option to renew.

Poland — the term is governed by the Polish Civil Code. Commercial premises leases are limited to ten years. A longer lease is treated as a lease for an indefinite term determinable by notice. Attempts have been made to grant longer terms by the use of reversionary leases. However, the Code makes attempts to circumvent it void and it is not yet clear whether the reversionary lease route works.

Rent review

France — offers a range of possibilities for rent review. There are statutory upwards or downwards reviews every three years if the local market moves by more than 10% in either direction, measured against a government index based on construction costs. On top of this there is annual indexation by reference to the same index. The complexity that this brings has led to the appointment of specialist judges to deal with disputes.

The Netherlands — for article 290 leases, yearly indexation in line with the equivalent of RPI is usual. The parties can apply to the court if the lease is extended by law or the tenant exercises its option to renew. The court can order a higher or lower rent. The same regime applies to article 230a leases except that the parties are not able to apply to court.

Hungary — annual indexation by reference to currency inflation is usual. Rent is sometimes denominated in Euros where an EU index will apply or US$ where the US consumer price index is used.

Poland — as most leases are between three and seven years long an annual index is used. As with Hungary, the index chosen will reflect the denomination of the rent.

Break options

France — only the tenant may have a break option

The Netherlands — both the landlord and the tenant will have a break right at the expiry of the every fifth year of the term. Article 290 leases have only a restricted right for the landlord to terminate during the first ten years. Article 230a leases enable a tenant who receives a break notice to seek an extension in the courts but this is rarely successful for more than a short period.

Hungary — the landlord may determine a lease not for a fixed term on not less than one year's notice. The tenant's corresponding right enables it to terminate at the end of any month on not less than a fortnight's notice. These arrangements can be varied by contract, usually to produce a compromise notice period of six or nine months for either party.

Poland — any arrangement is possible by contract, but the governing law enables either party to determine a lease for an indefinite term on notice.

Insurance

France — as in England and Wales, the landlord insures against damage by a full range of risks plus up to two years' loss of rent, at the tenant's cost. It is tax-efficient for the payment to be reserved as rent.

The Netherlands — as in England and Wales, the landlord insures against damage by a full range of risks and loss of rent. It is unusual for the tenant to reimburse the basic premium, but any additional premium payable as a result of the tenant's specific use and occupation of the premises will be repayable.

Hungary — the landlord will insure against damage by a full range of risks. The general law provides that destruction of premises terminates the lease so loss of rent insurance is unusual. It is, however, possible to contract out of this provision and pass all insurance costs on to the tenant.

Poland — while landlords usually insure multi-let estates or buildings themselves and recharge the tenants through the service charge, single-let premises are usually the responsibility of the tenant who, as in England and Wales, will be required to insure in the joint names of the landlord and tenant.

Tenant's repairing obligation (either directly or through a service charge)

France — the Civil Code will require the landlord to repair and maintain the roof, structure and foundations and the landlord will often be responsible for other repairs. Increasingly landlords are passing on the cost not only of non-structural but also of structural repairs to the tenant through a service charge.

The Netherlands — the law provides that the landlord will maintain the property at its cost. This obligation extends to structure, service media and external decoration. The tenant is responsible for internal and non-structural repair and is expected to act in a tenant-like manner in relation to those items which are the landlord's responsibility. It is possible for a tenant to be put on a full repairing liability, but this is rarely agreed.

Hungary — tenants assume responsibility for repairs of single-let premises. Landlords will maintain multi-let estates or buildings and seek to recover the costs through the service charge.

Poland — the starting position under the Civil Code is that the tenant is responsible for maintenance of the premises but the landlord will be responsible for the structural elements of a multi-let estate or building. The landlord usually seeks to recover these costs through the service charge.

Dealings with the landlord's reversion

France — an incoming landlord will assume all the rights and obligations under a lease which has been notarised or in respect of which necessary formalities with the tax authorities have been observed.

The Netherlands — all rights and obligations under the lease pass on a transfer. Where a sale is by a mortgagee the purchaser may have the option to end the lease.

Hungary — the original landlord's obligations continue after it has transferred its interest.

Poland — all rights and obligations under the lease pass on transfer. The incoming landlord will be able to determine a periodic tenancy and may be able to determine a fixed term tenancy unless notarising or registration formalities have been observed by the tenant.

Dealings with the lease

France — the Civil Code entitles the tenant to transfer its lease to a purchaser of its business. Other transfers require the consent of the landlord and the outgoing tenant must guarantee the lease obligations of its assignee. These provisions do not apply where there is a change of control of a corporate tenant.

The Netherlands — Article 290 leases may be transferred with the landlord's consent not to be unreasonably withheld. If the landlord objects the tenant can apply to the court for consent. This is available if the tenant's business is being transferred, the tenant will benefit financially from the transfer and the assignee is financially equipped (often by means of the production of a guarantee) to meet the lease obligations. The tenant may not have recourse to the courts to obtain consent for assignment of article 230a leases, however.

Hungary — assignment is rare as the concept of assignment is not contemplated by the Civil Code. While the landlord might agree to an assignment, it is entirely within his discretion. Landlords usually prefer to enter into a new lease with an incoming tenant. Outgoing tenants prefer to deal with their successors to the premises by granting an underlease, for which the landlord's qualified consent will usually be required.

Poland — any assignment requires the landlord's consent.

Alterations

France — structural alterations require landlord's consent. Non-structural alterations will either be permitted or will require the landlord's qualified consent. Commonly the landlord will require the right to supervise the tenant's works at the tenant's cost. The landlord will usually require the tenant to reinstate at the end of the term. The lease will usually prevent any claim by the tenant for the cost of improvements effected by the tenant.

The Netherlands — qualified consent will be required for any alterations which would cost money to reinstate at the end of the term. Tenants of article 290 leases can apply to the court for consent if the alteration is an improvement from the tenant's perspective and the landlord's consent is being unreasonably withheld. Tenants are usually required to reinstate the premises at the end of the term.

Hungary — structural alterations will require the landlord's consent.

Poland — structural alterations are permitted at the landlord's discretion. Other alterations are permitted with the landlord's qualified consent. In the absence of a contrary provision in the lease, the Civil Code gives the landlord the option at the end of the lease of "buying" the tenant's alterations or requiring their reinstatement.

US lease terms

A wide range of lease terms is available in the US. Notably, the US landlord is not fixated with the idea of a "clear" lease and there is no equivalent of the Landlord and Tenant Act 1954. Depending on the kind of property to be let, a tenant may choose from the following:

- Gross lease — This is the most commonly used form of lease, often used for office premises. The tenant pays rent; the landlord pays the outgoings (insurance, taxes and maintenance costs) relating to the property. Recently, gross leases have started to contain "escalation" clauses (see below). These enable the amount of rent to be adjusted (usually annually) to offset anticipated increases in the landlord's expenses.

- Net lease — This transfers some or all of the expenses that the landlord is traditionally responsible for under a gross lease to the tenant. With a "single net" lease, the tenant pays rent plus taxes relating to the tenant's premises. Under a "double net" lease, the tenant also pays the share of the insurance cost attributable to its premises. Finally, with a "triple net" lease (usually found in the context of large occupiers of significant premises), the tenant pays all charges payable under a double net lease, plus maintenance expenses. The triple net lease is analogous to a UK institutional lease and can be used by the landlord as a component of the securitisation process. (The detail of securitisation is beyond the scope of this book but can be shortly explained as the exchange of a future income stream for a present capital sum.)

- Fixed lease — This provides for a fixed rent over a fixed term. This type of lease is favoured by the small business tenant probably leasing smaller premises. There is no obligation to pay rent increases but the strategy is not without risk. If the tenant wants to renew the lease on expiry, the landlord may increase the rent. The increase can be steep if the tenant's business is successful and would suffer from relocation — as noted above, the absence of a 1954 Act-style regime can affect the tenant adversely.

- Step lease — This form of lease provides for predetermined rent increases at stated times. This offers some comfort to the tenant as it knows what its rental commitment on the premises will be, while giving the landlord some protection against cost inflation. The tenant should consider whether the scheduled increases are proportionate to the historic movement of consumer price indexes or local evidence of growth in market rents.

- Turnover or percentage lease — This lease entitles the landlord to share in the tenant's business success (or failure). The lease provides for a base rent, plus an additional amount that is set as a percentage of the tenant's turnover. This sort of lease is usually appropriate for retail premises.

- Escalation clause — This clause provides for increases in rent over a specified time period and may be compared with a UK rent review clause. The escalation can be fixed or linked to an external circumstance. This might be an increase in the landlord's operating costs, indexation by reference to an established set of statistics, or increases in the tenant's turnover.

- Maintenance — Responsibility for what would be service charge items in the UK is sometimes passed to the tenant. If it is, the lease will dictate whether the tenant can contract with anyone of its choosing to provide these services, or whether the service providers have to be approved by the landlord.

US REITS

REITs were created by the United States Congress in 1960 to facilitate investment in the kind of buildings and large-scale developments that would otherwise be inaccessible to small-scale investors and individuals. A REIT works by pooling the investors' capital into a single economic pursuit geared to the production of income through ownership of commercial property. REITs offer dual advantages to smaller investors:

- the risk-spreading advantage of investing in a portfolio of properties rather than a single building
- expert management by experienced commercial property professionals.

REITs may be generalist or specialise in particular sectors (offices, industrial, retail) or even sub-sectors such as car dealerships.

A US REIT must comply with four fundamental requirements:

(1) it must distribute annually at least 90% of its taxable ordinary income to shareholders
(2) most of its assets must be property related (this may include investments in mortgage loans)
(3) it must derive most of its income from property held for the long term
(4) its shares must be widely held.

By complying with these requirements, a REIT benefits from relief on dividends paid so that most, if not all, of its income is taxed only at the shareholder level. This tax transparency allows investors — including those outside the US — to receive a healthy dividend yield.

The other side of the coin is that a REIT may retain only a small proportion of its income. As a result, capital for acquisitions, refurbishment and redevelopment comes mostly from funds raised in the investment marketplace, sale proceeds and joint ventures with specialist concerns.

When a REIT sells property in a taxable transaction, it may either distribute the proceeds as a capital gain dividend (which also qualifies for a dividends paid deduction against a corporate-level tax) or retain them and pay tax at corporate level on the gain. It is unusual for a REIT to choose the second option.

Since 1991, the growth of REITs has facilitated increasing ownership of institutional quality commercial property in all sectors. Statistics produced by Prudential Real Estate Investors show that office REITs own 7.9% of office space, industrial REITs own 8.5% of industrial property, retail REITs (excluding shopping centres or "malls" as they are known in the US) own 13.1% of the retail market; and mall REITs own a remarkable 36.8% of malls.

A brief case study — the industrial sector in the US

Moody's, the rating agency, has designated industrial REITs are one of their highest-rated REIT sectors. There are a number of reasons for this:

- industrial REITs offer high levels of cover for fixed and interest charges
- most REITs in the sector have relatively low levels of secured borrowing
- the industrial sector is not particularly vulnerable to overbuilding because of short construction periods and a preference for "turney" development
- an industrial tenant will often renew its lease because it usually has fixed equipment at the premises that is expensive and difficult to move
- stable cash flow is an inherent credit strength of the industrial REIT sector. A lease of a new industrial building will have a seven- or ten-year term. Average lease lengths in the industrial sector are at the long end of the range, a position shared with the office and regional mall sectors. (This point is particularly interesting from a UK perspective; both the initial and average lease lengths look low by our "institutional" standards.)

- industrial REITs have the advantages of economies of scale and specialisation that enable them to concentrate on high-quality tenants and those with "bespoke" requirements

The shift from feudalism to consumerism is marked in the US industrial sector. Changing tenant requirements have forced many industrial landlords to place greater emphasis on customer relationships. This is a sea change in the landlord's *modus operandi*. The commercial property business has historically offered an inflexible product. Tenants now require a product that suits them both in terms of location and price. Competition in the sector dictates that the landlord offers not just a building but also a wide range of "value-added" services to meet its tenants' needs. The quality of service and its delivery are also crucial. Some landlords have established strategic partnerships to offer their customers — as one must learn to think of the tenants in this context — a wide selection of additional services ranging from facility design and procurement to the design of computer software applications tailored to the tenant's business.

The landlord now employs executives with business as well as property experience and expertise. It is thought that the successful anticipation of the customer's needs will be the difference between a landlord that forges ahead and one that simply survives. A landlord who is successful in this respect can expect to retain tenants longer and extract above-market rents. This virtuous circle saves marketing costs, strengthens the landlord's financial position and enhances the landlord's reputation, leading to further success. We hear that some US REITs already make more profit from the additional services they sell to their customers/tenants than they do from traditional rental income.

UK REITs

At the time of writing the UK Government has just announced that REITs can be established in the UK from 1 January 2007. This decision followed a lengthy consultation period during which the UK property industry moved from gloomy prognosis about the Government's proposals to grudging acceptance of them. Indeed, a number of significant listed property companies, including British Land, Slough Estates and Brixton have stated their intention to convert to REIT status in 2007. As so often before, where the US leads, the UK follows. It remains to be seen whether this will be the end of the institutional lease as we know it.

Acknowledgements and Further Reading

Books and other publications

A Code of Practice for Commercial Leases in England and Wales, RICS Policy Unit March 2002

A Guide to the Changes to the Landlord and Tenant Act 1954 Dechert LLP June 2004

"A New Lease of Life" Crosby, Neil; Murdoch, Sandi; Hughes, Cathy, *Estates Gazette* 5 March 2005

BPF and BCO Model Terms for an FRI Office Lease of Whole February 2003

Cushman & Wakefield Healey & Baker Corporate Briefing on the New Accountancy Rules C&W H&B April 2004

Handbook of Rent Review Reynolds QC, Kirk and Fetherstonhaugh, Guy, Sweet & Maxwell 1981

Hill & Redman's Law of Landlord and Tenant Barnes, Michael; Matthews, Paul; Harry, Timothy; Taggart, Nicholas; Moss, Joanne R; Furber, Joanne R, Butterworths Law 1996

"Law's the Pity" Romney, Charles, *Property Week* 10 May 2000

Leasehold Liability: Landlord and Tenant (Covenants) Act 1995 Fogel, Steven; Riley, Alan; Rogers, Philip; Slessenger; Emma, Jordans 2000

Rent Reviews and Variable Rents Clarke, David and Adams; John, Oyez

"Stamp Duty Land Tax" Boutell, Mike; *The Tax Journal* 24 May 2004

The Landlord and Tenant Factbook Marsh, Matthew; Bhaloo, Zia; Members of Collyer-Bristow and Enterprise Chambers, Dechert LLP, Sweet & Maxwell 2000

The Strutt & Parker/Investment Property Databank Lease Events Survey 2005

"Understanding French Lease Law" Knopp, Corinne and Varlet, Caroline *Property Law Journal* 15 May 2006
Woodfall: Landlord and Tenant Brock, Jonathan; Dowding, Nicholas; Morgan, Paul; Rodger Martin, Sweet & Maxwell 1978

Organisations

Investment Property Databank (*www.ipdindex.co.uk*)
Practical Law Company (*www.practicallaw.com*)
Taxation Web (*www.taxationweb.co.uk*)
Accounting Web (*www.accountingweb.co.uk*)

And thanks to...

Malcolm Dowden of Charles Russell (whose idea this book was in the first place)
Emma Slessenger of Allen & Overy
Simon Latham of ING Real Estate Investment Management
Ed Saunders of Cushman & Wakefield Healey & Baker
Professor Neil Crosby of the University of Reading

Index